SUBURBAN FIRE TACTICS

SUBURBAN FIRE TACTICS

JIM SILVERNAIL

Fire Engineering®

Disclaimer

The recommendations, advice, descriptions, and the methods in this book are presented solely for educational purposes. The author and publisher assume no liability whatsoever for any loss or damage that results from the use of any of the material in this book. Use of the material in this book is solely at the risk of the user.

PennWell Corporation
1421 South Sheridan Road
Tulsa, Oklahoma 74112-6600 USA

800.752.9764
+1.918.831.9421
sales@pennwell.com
www.Fire EngineeringBooks.com
www.pennwellbooks.com
www.pennwell.com

Marketing Manager: Amanda Alvarez
National Account Manager: Cindy J. Huse

Director: Mary McGee
Managing Editor: Marla Patterson
Production Manager: Sheila Brock
Production Editor: Tony Quinn
Book Designer: Susan E. Ormston
Cover Designer: Karla Pfeifer

Library of Congress Cataloging-in-Publication Data
Silvernail, Jim.
Suburban fire tactics / Jim Silvernail.
pages cm
Includes index.
ISBN 978-1-59370-294-6
1. Fire extinction--United States. 2. Suburbs--United States. I. Title.
TH9503.S45 2013
628.9'25--dc23
2012046761

Printed in the United States of America

1 2 3 4 5 17 16 15 14 13

This book is dedicated to my father, retired Chief James Silvernail.
I am the son of a fireman. For many outside of our profession,
it might seem disrespectful to make this comment considering
the fact that my father served as a fire chief for more than 30 years.
However, to me it is the greatest sign of respect.

Contents

Part III: Formulating a Suburban Fire Tactics Game Plan

Preface

Suburban Fire Tactics is a guide for suburban-based fire operations. All fire departments in the country share the same common objectives: to save lives, protect property, and to limit harmful impacts to the environment. The only differences between urban, rural, and suburban agencies are in the delivery of tactics and strategies.

Often, suburban-based fire departments fall into a trap by attempting to use the same tactics as large urban city fire departments. Unfortunately, this is not always safe or successful. *Suburban Fire Tactics* explains the factors that limit suburban operations and the differences between urban and suburban capabilities. It provides the suburban policy makers with the means to create and tailor suggested operating procedures for their specific circumstances and discusses how to prioritize and conduct fireground functions related to the suburban situation. *Suburban Fire Tactics* also illustrates some application guidelines developed for suburban firefighting scenarios.

Suburban Fire Tactics is written with the main goal of assisting suburban-based fire agencies in the delivery of fire suppression services and serving as a *tactics and strategies* manual. The intended audience is broad and ranges from policy makers and chief officers to company officers and firefighters. It provides several tools:

- A templated, systematic approach for the strategic planning of suburban-based structural firefighting
- A guide for fireground decision making
- A collection of practical applications used within the fire service industry for suburban fire operations

A suburban agency attempting to accomplish the same tactics and functions as its urban counterpart can be a formula for catastrophic failure and unsafe operating practices. Many times, these organizations will dodge a bullet and accomplish their objectives on the fireground. Too often, however, an effective standard operations procedure designed for an urban environment that is tested in a suburban environment fails due to poor situational awareness and a misjudgment of capabilities.

To illustrate, the book shows examples where suburban-based operations have faced failure and ineffectiveness when utilizing the wrong tactics. One such common example of this problem is the failure to reach the main body of fire with the initial attack line. We analyze the reasons for some actual failures in part II of this book, but there are some general tactical errors that can contribute to a botched outcome:

- Inadequate size-up/inability to quickly locate the body of fire
- Inadequate staffing for an effective hose stretch
- Inadequate ventilation
- Inadequate forcible entry

Suburban Fire Tactics provides a systematic approach to correcting these inadequacies and provides a framework for safety and suburban fireground success.

How to Use the Manual

Suburban Fire Tactics is divided into three parts. Part I begins with the mission of the fire service and explains the reason for its existence. It explores the concepts and objectives of firefighting, and discusses fire behavior and fire attack principles. Chapter 1 discusses the systematic approach to developing a plan for the implementation of suggested operating procedures (SOPs). These SOPs are the answer to solving the challenge to "different tactics." It explores the concepts and objectives of firefighting and discusses fire behavior and fire attack principles. It concludes with the challenge for suburban firefighting: "same objectives, different tactics." Chapter 2 evaluates the differences between urban and suburban operations. This chapter covers prioritizing fireground functions and actions, developing SOPs, planning implementation strategies, and making plan reassessments. This section is written with the intent to assist the policy maker with SOP development and to assist company officers with prioritizing fireground functions.

Part II discusses the specific application of suburban firefighting operations. Chapters 3 through 10 explain how to conduct an accurate size-up, assist in initial decision making as a first due company officer, and provide solutions for water supply challenges, commercial occupancies, and technical rescue requirements.

Part III provides information needed to create the game plan, including tactical variables, implementation suggestions, and preevaluation and planning guidelines.

Systematic Approach

The systematic approach utilizes a strategic planning model developed for specific service delivery (fig. P–1). The process ultimately links all actions taken at the fireground back to the mission of the fire service in order to achieve desired outcomes. Through this model, the policy maker delivers expected services and guarantees a continuity of operations with an emphasis toward firefighter safety. Haphazardly operating a fire service organization without a plan is not responsible and can result in disaster. Those who do not plan simply react to problems. Visionaries who plan are proactive and create a desired result. These visionaries are true leaders who have accepted responsibility for their organization and the welfare of all stakeholders.

Fig. P–1. A systematic, strategic planning approach

Of course this all looks very academic and the model may seem out of place in a tactics and strategies manual. The reasoning behind this is to understand the methodology behind deriving "why we do what we do" and staying on target. You aren't going to find this applied strategic planning model in an SOP manual. However, as a policy maker, it is essential that you approach your responsibilities with a systematic method, specifically strategic planning.

Planning to plan, values scan, and mission formulation

For the majority of us, the first components of the model have already been completed. The *planning to plan*, *values scan*, and *mission formulation* has been completed by organizational predecessors and community leaders. Generally, our existence has been predetermined by the fire service industry and the communities we have chosen to protect.

Strategic business modeling

In relation to developing suggested operating procedures, the big challenge for policy makers falls within the next component: *strategic business modeling*. In relation to structure fires, strategic business modeling is equivalent to defining which objectives we are trying to obtain.

Performance audit and gap analysis

This text focuses heavily upon *performance audit* and *gap analysis,* as they are the crucial elements of this model. It is the difference between urban and suburban service delivery. The performance audit and gap analysis, if completed accurately, will realistically indicate if objectives can be achieved with current resources.

Integrating action plans

Action plans in the form of suggested operating procedures or suggested operating procedures will be developed to close the gaps and achieve objectives.

Contingency planning

"Preferred" and "suggested" are important concepts that allow for the *contingency plans* to also be developed. If the initial preferred plan doesn't work, a backup plan will have been created to handle unforeseen circumstances.

Implementation

Finally, the plan can be implemented. For the purposes of structure fires, this includes the coordination of fireground activities that have been enacted to achieve tactics and strategies.

Environmental scanning and application considerations

Environmental scanning and *application considerations* are continuously monitored from the inception of the plan through implementation. Once again, environmental and application considerations are major differences between suburban- and urban-based operations. A systematic approach also includes the reassessment of the implementation for continued effectiveness.

For simplification purposes and easy use of *Suburban Fire Tactics*, the following chart identifies key structural firefighting concepts associated with each tactical component and in which chapter they are discussed.

Component	Concept/Action	Chapter
Planning to plan	Identifying a need	Ch. 1
Values scan	Understanding attitudes, accepted fire service principles	Ch. 1
Mission formulation	Defining mission/goal, fire service concepts and priorities	Ch. 1
Strategic business modeling	Planning fire service objectives	Ch. 1
Performance audit/gap analysis	Understanding specific fire agency capabilities	Ch. 2
Integrating action plans	Developing SOPs	Ch. 3
Contingency planning	Developing alternative plans of attack	Ch. 3
Implementation	Supporting, conducting, and coordinating fireground functions	Ch. 12
Environmental scanning	Defining response area characteristics, training situational awareness, reevaluating	Ch. 11 Ch. 5 (reevaluation)
Application considerations	Understanding available resources, reevaluating factors	Ch. 11 Ch. 13 (reevaluation)

Conclusion

Let circumstance dictate actions. *Suburban Fire Tactics* illustrates some of the different circumstances suburbia presents and the coordination of efforts that must be performed for fireground success. The solution is to implement a systematic yet dynamic plan known as *suggested operating procedures*.

Acknowledgments

Suburban Fire Tactics is a compilation of experiences that my colleagues and I have collected over our careers. It began with the love and support of our families and grew with the shaping by key leaders. I have been very fortunate to have been trained and influenced by so many true firefighters throughout my journey in the fire service. I have also been blessed to work for an organization that has fostered a safe and progressive work environment, without which this project would not have been possible. I am very proud of the Metro West Fire Protection District and the support they have provided.

A book like this cannot be written by one person. It is a collection of concepts and expertise that are compiled by numerous industry experts. I would like to thank all of my contributing authors for their hard work and much needed additions to *SFT*: Steven Heidbreder, Michael Digman, Byron Long, John Shafer, and Scott Hulsey. Most importantly I would like to give credit to Erin McGruder, my project manager and personal editor. Erin is a fellow member of the plans team at Missouri Task Force-1 (USAR) and firefighter with Boone County (Missouri). Without Erin, I would not have had the discipline to fill in the blanks and organize my thoughts in a timely manner.

I have two families, as does every firefighter. The first family whose support I would like to acknowledge is my fire service family. Over the years I have encountered some of the most interesting and knowledgeable people who have ever walked the planet. I grew up in a firehouse. My fire department experience began as a volunteer/reserve firefighter with both the Metro West and Eureka Fire Protection Districts and quickly transitioned into a career when I was appointed to the Maplewood Fire Department as a firefighter and then to the Metro West Fire Protection District. I have made many friends along this journey that will always remain close to me.

Currently I serve as a company officer to one of the finest crews in suburbia, if not the United States. To my friends and fellow crew members Smitty, Michael, Scott (NN), Jeff, and Matt, thanks for putting up with me while writing this book. To Steve and John, my fellow captains and leaders, thank you for the ideas, camaraderie, and inspiration. Onward goes the Big Blue 4.

I must also give credit to two other fine organizations that I proudly serve: the St. Louis County Fire Academy and Missouri Task Force-1. I am proud to identify myself as an instructor, a title that has been bolstered by the St. Louis County Fire Academy. To my fellow instructors and former students, go forth with this job that we love so much. Finally, to my fellow members of Missouri Task Force-1, the pride that was felt by your efforts in New York on that September day cannot be topped.

Last but not least, thank you to my family. Without their love, sacrifice, and dedication, this book would not have been written. To my beautiful wife, Krista, and my four wonderful children, Alex, Sarah, Will, and Maura, thank you for being right next to me and being my support system. You are the reason for my determination. To my mom, dad, sister, grandparents, in-laws, uncles, aunts, nephews, nieces, and cousins, thank you for being you and having such an impact on my life. *Suburban Fire Tactics* is for all of us.

Part I

Introduction to the Suburban Fire Tactic Methodology

Fire departments throughout the world all exist with a common mission, regardless of the setting. Urban, suburban, and rural departments are all faced with the same basic fireground objectives: rescue, containment, confinement, and extinguishment. All agencies, however, are not created equal and have significant differing circumstances. These characteristics force each agency to find ways to obtain these established objectives with customized tactics, developed specifically for individual agency capabilities.

The first two chapters of this text, "The Challenge: Same Objectives, Different Tactics" and "Developing the Game Plan: Suggested Operating Guidelines" reveal the theme and true driving force behind *Suburban Fire Tactics*. The reality is simple: we have all have the same job to accomplish, but different parameters and circumstances under which to operate (*same objectives, different tactics*).

The *Suburban Fire Tactics* methodology contains two major segments: the challenge and the solution. The solution is devising a systematic game plan and implementing tactics that maximize resources and achieve the most efficient fireground procedures. For the fire service, this systematic game plan is a predeveloped set of procedures known as suggested operating guidelines.

Chapter 1

The Challenge: Same Objective, Different Tactics

The Challenge Illustrated

The journey begins with a dilemma. For more than 250 years, a special breed of citizen has answered the calling to the role of firefighter. This is not a new concept, nor is the idea of developing tactics and strategies for the purpose of combating structure fires. So, why the need for a text that primarily focuses on the tactics and strategies for suburban firefighting? The answer to this question is capability. Capabilities for each individual agency across the globe differ depending on fundamental variables including staffing, equipment, and infrastructure. Organizations, dependent on these variables, must devise customized tactics and strategies to combat structure fires and eliminate the dangerous *tradition-based* approaches that engage our operations. For safe and effective operations on the fireground, suburban- and rural-based operations cannot utilize the same playbook as urban departments; the variables are way too different.

Let's begin by looking at an example:

> *It's 3:00 on a warm August morning. Temperatures have been over 100°F all week. Luckily it's early morning, yet temperatures are still in the 90s. You are awoken for a report of a structure fire in an apartment complex. Quickly, you and your crew don your gear, board your apparatus, and take off for the reported location. En route, you are notified there may be numerous people trapped in the interior and on balconies. Because you are a good officer, you have preplanned this complex thoroughly and know it is*

occupied by many elderly residents with mobility issues that complicate life hazards.

Upon arrival you find a scene of chaos at a three-story, garden style apartment building with 12 separate units. Access is limited by double-parked cars throughout the complex. Heavy fire is visible from two windows on the second floor. A 360 degree size-up finds four people on third floor balconies on the C side and one resident who is badly burned lying in the grass on the A side. Smoke under pressure is also visible pouring from the interior of the stairwell, indicating an open door to the fire apartment.

This picture sounds like a company or chief officer's worst nightmare. Is it real? It absolutely is real and I recently experienced this challenge. Is it time to panic and give up? No! We have a responsibility to mitigate the loss, regardless of the situation. How does your agency go about organizing this chaos and making the most effective impact to this situation? It begins with a game plan and developing operating guidelines that work for your specific agency (fig. 1–1).

Fig. 1–1. Planning and operating guidelines can help organize chaos and provide the most effective impact to a situation. (Photo provided by Southern Stone Fire Protection District)

Why specific? Won't a generic suggested operating guideline (SOG) work for every organization? To answer this, consider the typical urban response to a structure fire, thoroughly analyzing critical variables.

> *You are the company officer of an urban engine. Unfortunately, you have to wait your turn down the sliding pole because the ladder company is also detailed to this assignment. A total of nine people are scurrying for their positions on their apparatus. You hop in the front seat, and turn around to make sure that all three firefighters behind you have geared up and are fastened in. The engine takes off and you take a look at your computer-aided dispatch terminal to see what engines are also assigned. You feel a sense of urgency because you do not want the neighboring engine to beat you, as first due. The screen also shows you there are hydrants on every corner, approximately 400 to 600 ft apart.*
>
> *The engine arrives and you know that your first priority is to get control of the situation (fig. 1–2). Because of the occupants, you elect for an aggressive offensive/interior stretch. You coordinate with your truck company, which takes fire floor truck operations while you stretch your line and find the main body of fire. You extinguish the fire and the truck makes the quick rescues. The district chief arrives quickly and takes command of the incident. The second arriving engine that almost met you at the intersection stops at the hydrant behind you, forwards a large-diameter supply line, and then quickly deploys a secondary attack line.*

Now let's look at this situation from a suburban perspective.

> *You are the suburban company officer. You board the apparatus and turn around to make sure that your only firefighter behind you is on board. You then look at the computer to see that the next arriving apparatus will be lucky to be there within 5 minutes of your arrival and that there is only one hydrant, approximately 500 ft from the reported fire building.*

Fig. 1–2. Flames are showing from the second story of an apartment complex. (Photo provided by Southern Stone Fire Protection District)

> *You arrive and need to make an immediate decision: Do I focus on rescue or do I try to control/confine the fire? Can I effectively place a line on the second floor of adequate size in a timely manner to control the fire with just two personnel? How much water do I have? How do I effectively manage this scene?*

With a quick glance at both scenarios the reader can immediately observe the contrasting variables that will affect operations:

- Assigned/available apparatus (including functional apparatus: trucks vs. engines)
- Assigned personnel
- Water supply

Of course this is should be obvious, but what do these variables truly affect?

- Initial mode decision making (the possibility of viable rescues always requires an offensive mode)
- Initial actions (immediate rescue vs. line deployment)
- Water supply considerations (forward, reverse, delaying for second engine)

- Assuming command (Who initially assumes command? How do you interpret Incident Command System (ICS) and National Incident Management System (NIMS)?)
- A safety net for operations (Who deploys backup lines? Who assumes rapid intervention team (RIT) responsibilities?)
- Essential truck functions (Who conducts essential truck functions if you do not have truck companies? What does rescue look like? Who does rescue? Can you perform all forms of desired ventilation effectively?)

This simple example alone illustrates the need and idea behind *Suburban Fire Tactics*. This text provides examples of how to organize the chaos of a structure fire and how to customize your individualized response plans, considering the basic elements/functions of modern day firefighting.

The Mission

Fire agencies all over the world, regardless of their organizational structures or location, share a set of common goals. These goals, or mission, are the reason for the fire service's existence. Fire departments, districts, and authorities are all created out of a community's basic necessity for life safety and property protection. It does not matter if the organization is urban, rural, or suburban as this focus is the same throughout. Although the delivery methods may differ, the basic desired outcomes all look the same.

Since the existence of early civilizations, mankind has understood a need for protection against the dangers associated with fire—a major threat to human life and property. Hand operated water pumping devices were first documented in ancient Egypt. In Rome around 80 BC, Marcus Licinius Crassus created the first bucket brigade to protect against developing conflagrations. Although he developed the service for his own personal gains, it demonstrated the realized need for protection existed. In AD 6, Emperor Caesar Augustus organized "Vigiles," a bucket brigade intended for public fire protection. The first documented fire engine company in the United States was established January 27, 1678. Even the famous American inventor Ben Franklin had his hand in the fire service. In 1736 Philadelphia, Franklin established the Union Fire Company. Cincinnati, Ohio, formed the first fully paid fire department in the United States, getting its start April 1, 1953.

Before tactics and strategies can even be considered or developed, the mission of a fire service organization must be understood. There are three basic principles:

- To save lives
- To protect property
- To reduce hazardous impact to the environment

Each organization may word it differently, but these are the reasons firefighters all over the world conduct their operations and assume the responsibility of maintaining a fire service agency. It is a uniformly adopted creed that should not be forgotten.

As an industry we also recognize a need to prioritize our goals and rank them in a hierarchy of importance. Many may say it is common sense to rate human life higher than that of property. It is common sense, but as fellow human beings, it is also a moral and ethical responsibility. Even so, organizations can lose sight of this principle. Perhaps not intentionally, but often tactics can be initiated with blinders that place an emphasis on property conservation or have an unnecessary disregard for the firefighter's life safety. Yes, firefighters do pledge to risk their lives to save their fellow man. What is completely unacceptable is to risk lives over property. Therefore, we must focus on the mission. Later in this text we discuss how to prioritize objectives, tactics, and functions to correspond with and support this priority.

Mission Statements

When discussing tactics and strategies it is essential to relate all job roles, tasks, and functions back to the mission of the organization. At 3:00 a.m., when your company pulls up to a structure with fire blowing from two windows and the possibility of someone trapped, it is hard to believe that a document titled "Mission Statement" can have any value to your immediate operation. But it does! It gives you, as a firefighter, a reason for being. It gives direction and points out what is important, who is important, and each person's role (fig. 1–3). Of course you don't pull up and recite by memory the entire mission, which was probably created by someone working from a nice safe office. However, all your actions, values, and priorities have been predicated and impressed upon you through this simple document.

Fig. 1–3. A mission statement is the foundation for on-scene decision making: what is important, who is important, and your role. (Photo by Box Alarm Production Fire Ground Photography)

Effective mission statements have common elements. They do not need to be complicated, yet they need to be able to convey the direction and goals of the organization. The minimal elements should include

- The function of the organization (what the organization does and the services it provides)
- The stakeholders in the organization (everyone involved, both internal and external)
- How the organization supplies the services
- Why the organization exists

An example of a good mission statement could like something like this:

> *"The XYZ Fire Department is dedicated to protecting the lives and property of the citizens of XYZ City by providing the best services through the finest personnel, vehicles, equipment, and training that can be provided within the constraints of the fiscal consideration and wishes of the citizens."*

This is a simple statement that identifies what, who, how, and why.

- What: dedicated to protecting the lives and property of the citizens of XYZ City
- Who: the stakeholders are the XYZ Fire Department, citizens of XYZ City, personnel
- How: finest personnel, vehicles, equipment, and training
- Why: the wishes of the citizens

The mission statement's purpose is to give direction and link all organizational activities toward the shared, pre-established common goals. In the corporate world, the mission statement keeps a company from diversifying away from its main reason for being in business. It keeps the company focused. In many instances, entities that stray too far from their main path often fail because their intended business models will not support this diversification. In the fire service, we aren't making products. The services we produce are extremely sensitive; sensitive in the meaning that if incorrectly or not provided could make the difference between life and death. Therefore, if services, values, and objectives aren't supporting the mission of the organization, they need to be either corrected or eliminated.

Note: The trend in industry and business is to simplify the mission statement in order to not clutter or confuse the main point for existence. Many agencies have chosen to represent their purpose with one strong catch phrase or sentence and have included all technical information in the vision component of the organization. Regardless, all agencies must have a concise *purpose* that is understood throughout.

Priorities: The Reason for Existing

Once the importance of the mission statement is understood, it is essential to further detail the three main reasons for which the fire service exists: to save lives, to protect property, and to minimize harmful impact to the environment.

Saving lives

According to the National Fire Protection Association (NFPA), there were 3,120 civilian deaths and 17,720 civilian injuries caused by fire in 2010.

The 482,000 structure fires in that year accounted for 2,745 of the civilian deaths and all 17,720 civilian injuries (fig. 1–4). One civilian death occurred every 169 minutes, while a civilian injury occurred every 30 minutes in the United States. From 2005 to 2009, a yearly average of 2,881 civilian deaths and 14,945 civilian injuries were caused by structure fires.

Fig. 1–4. The 482,000 structure fires in 2010 accounted for 2,745 civilian deaths and 17,720 civilian injuries. (Photo by 5280 Fire)

Fire has always been seen as a constant enemy and a hazard to life. Communities have recognized a need for protection against this threat and found a remedy in the form of fire protection. Firefighters are all made aware, from the moment they enter their first day at the fire academy, that firefighting is dangerous and that they may be asked to risk their lives for another human being at some point in their careers. Therefore, the credo of the fire service begins: "We risk a lot, to save a lot." What constitutes a lot? Is there anything more precious than human life? The answer is a resounding *no*. Because we do not send robots to rescue people from burning buildings, "a lot" also signifies the human resource used. Therefore, in this context, substitute "human lives" for "a lot."

What if there are multiple lives at stake? Since price cannot be placed on human life, can we place importance of one individual over another in life saving decision making? The answer once again, for both moral and ethical reasons, is *absolutely not*. Therefore, the fire service has another credo when dealing with life safety: "Do the most to impact the most." In other words, firefighters will do the most that is possible to help or save the maximum number of people possible. Firefighters will try to initiate tactics that will help impact the whole, not actions that will benefit the few. This is the model for life safety efficiency.

If fire crews are unable to reasonably rescue everyone in the same time frame due to resources available, they will begin to focus attention on those immediately in harm's way and to prioritize from this position. In this case, firefighters will make decisions without discriminating and without bias.

The rules for saving life are

- Risk a lot to save a lot
- Do the most to impact the most
- Focus on those closest to harm's way, without bias

Protecting property

Firefighters across the world have sworn an oath to protect the property and assets of the citizens they protect. To what extreme do fire organizations commit in order to assure this guarantee? At no point is risk of property loss worth more than risk to life, including those of firefighters. However, does that mandate no offensive interior operations should be undertaken on vacant structures? Before these questions can be answered, we must define the meaning of "risk" and be able to identify "vacant."

Risk assessment. It is an industry standard that we accept a small risk when tasked with saving property. Hence, "we risk a little to save a little." That is a very broad concept. To understand the true meaning of this saying, risk must be understood. The hazards of structural firefighting are a threat to every firefighter. The risks we face can be minimized through experience, training, proper equipment, and safe operating practices. Therefore, the risks assumed by each agency are not the same and should be assessed differently.

Occupational risks can be categorized into four groups according to their severity and frequency:

- Risks of a high frequency and low severity of impact/damage
- Risks of a high frequency and high severity of impact/damage
- Risks of a low frequency and low severity of impact/damage
- Risks of a low frequency and high severity of impact/damage

Why are we concerned with categorizing risk? Shouldn't we be concerned with all risks? Of course the answer is *yes!* In order to minimize risk, though, we must truly understand its impacts. Generally, in any occupation, minimizing the impact requires adopting policies and standards to overcome the risk. Risks that are experienced on a regular basis are generally understood and anticipated. Experience is golden. Another generalization that can be made is that severity also gets attention. If an event is perceived as severe, precautions are taken or unsafe acts are avoided.

Across the country, the act of fighting structure fires commonly falls within the category of low frequency/high severity. This by far is the most dangerous of all four for several reasons. Most firefighters like to consider themselves experienced professionals. The fact of the matter is, because fires generally don't happen every day, especially in suburban areas, it takes many years to build competencies. Also, structure fires have too many variables and each fire scene is different. Most importantly, the hazards associated with structural firefighting can be severe and lethal. In 2010, the United States Fire Administration reported 87 in the line-of-duty deaths associated with structural firefighting (fig. 1–5).

Once the nature of risk associated with structural firefighting is understood, organizations need to find ways to assess and minimize risks in order to accept the responsibility of initiating offensive firefighting operations to protect property and assets. First, all agencies must provide all essential personal with protective equipment and strictly enforce its use. Secondly, through industry and organizational experiences, policies and operating procedures must be developed that promote safety. The development of these procedures is further discussed in later chapters of this text. Finally, the organization has to proactively shift the functions associated with structural firefighting from *low frequency* to *high frequency*. This can be artificially accomplished through training.

Unfortunately, not all fire service organizations are created equal. They may not have the resources to provide the essential equipment or have the ability to effectively train. They may also lack the experience, due to call volume or lack of effort, to institute policies or procedures. In these cases, the risk is much greater. The individuals in these organizations should realize their limitations and not assume the same.

Because the fire service understands risk assessment and minimization of risk, safe but aggressive strategies and tactics can be initiated to protect property. Aggressive, however, cannot be used without the key word *safe*. Those organizations that understand the importance of proper personal protective equipment, safe operating procedures, and training have minimized their risks and are compliant with "risk a little." Please note that a large component of safe operating procedures includes the effective structural size-up to determine scene stability and predict hostile fire events. Structural size-up is discussed in detail in chapter 3.

Fig. 1–5. Structure fires contributed to 87 line-of-duty deaths that occurred in 2010. (Photo by Box Alarm Production Fire Ground Photography)

Determining vacancy. The *unknown occupancy* is another important aspect of protecting property and assets (fig. 1–6). Determining if a life threat exists in a structure fire can only be determined by fire service crews. Every involved structure should be considered occupied until deemed vacant. Regardless of whether the structure is reported vacant or it is reported fully evacuated, primary searches or confirmations need to be completed.

Fig. 1–6. Occupancy clues, such as time of day or type of occupancy, should be noted during size-up. (Photo by Harry Hamm)

Often, vacant buildings become a sanctuary for the homeless. When people squat in these buildings, it is very difficult to determine their locations or last known whereabouts. It is also difficult to tell how many occupants may be residing illegally in the structure. An example of this was a fatality fire in New Orleans on December 28, 2010. Crews from the New Orleans Fire Department responded to a report of a fire in a vacant building on St. Ferdinand and Prieur at approximately 0200 hours. Crews found eight deceased homeless persons who had sought out the building for shelter.

When conducting size-ups of both residential and commercial buildings, there are clues to look for to determine possible occupancy:

- **Time of day.** During the night there is a higher occupancy possibility in residential buildings; during the day there is a higher possibility for occupancy in commercial buildings.
- **Type of occupancy.** The type of occupancy should be an indication about the possible number of occupants. For example, an apartment complex during the evening, or a school during hours of operation, would indicate a high target (life) hazard, requiring more resources than would a single-family residential structure.
- **Vehicles present in front of the structure**
- **Bystander accounts.** Even though they are not always accurate, bystander accounts can be a quick, initial account of the whereabouts of exact location of occupants. It can be a direct pinpoint to victim location.
- **Board-up removed.** Vacant buildings that are not secured are targets for illegal occupation.
- **Unexplained fire origins in vacant structures.** Improvised heating sources may contribute to the origin of the fire. This was the case on December 3, 1999, in Worcester, Massachusetts, when homeless persons started a fire in a cold storage warehouse and six firefighters lost their lives searching for the victims' whereabouts.

The clues are not a 100% assurance that occupants are or aren't still inside the structure. If possible, primary searches should still be completed to verify occupancy.

Worth. What are properties or assets truly worth? Can we make assumptions of what is important or what is less valuable? The only thing that firefighters can make an assumption on is that life is by far more valuable than anything. When it comes to personal property, however, we cannot discriminate and assign our own value systems. Everything that was sworn to be protected should be considered of equal value. An individual who owns a $25,000 trailer home has the same right to service as the owner of a $1 million mansion. One person's perceived junk may be another's gold mine.

Never assume property insurance will take care of all the problems incurred by a fire. There are certain items that absolutely cannot be replaced, such as pictures, documents, and heirlooms. Money could never bring these objects back. Also, not every insurance policy is the same. In fact, the less fortunate may not have any property insurance. The important lesson to remember is that, as an industry, we do not discriminate property. We make all efforts to safely save and protect assets. It is not ethical for firefighters to judge worth or to discriminate.

Minimizing harmful impact to the environment

More importantly than ever, the fire service has recognized a third priority in public protection. Fire organizations have taken a responsibility to protect the environment. Therefore, not only are firefighters concerned with the effects of fires on people and property, but also the short- and long-term consequences to the Earth.

When arriving at structure fires, or any emergency, crews should consider possible harmful effects that the event can cause to the surrounding environment (fig. 1–7). Objectives should be initiated to limit and confine harmful impacts. Generally, in relation to detrimental impacts to the environment, the fire service concentrates on the possible presence of hazardous materials. Hazardous materials can come in the form of explosives, flammable and combustible substances, poisons, and radioactive materials. Hazard releases generally occur during transportation, production, storage, use, or disposal. Fire service crews must be proactive to identify where these agents are located within their venues and have a plan for containment and mitigation.

Today's modern fire service includes hazardous material incident mitigation as part of service delivery. Most fire agencies train personnel to the awareness and operational skill levels, and many areas have competent hazardous materials response teams that are trained in the containment and mitigation of hazardous materials. At the minimum, fire agencies should have in place a plan to contain any such hazards until trained professionals can be called to mitigate the situation.

Regardless of the training level, objectives should be established to minimize the harmful effects an incident may have on the environment. The key is to be proactive and identify these hazards prior to an emergency situation.

Fig. 1–7. Objectives should be established to minimize the harmful effects an incident may have on the environment. (Photo provided by Southern Stone Fire Protection District)

Fire Service Objectives

Missions are accomplished through strategic planning. Plans are enacted through the establishment of objectives. In the fire service, objectives are universal as well. The principles of fire protection are the same:

1. Locate
2. Rescue
3. Contain
4. Confine
5. Extinguish

These stated objectives are listed in an order and should follow a systematic progression. Extinguishing a fire is not as simple as haphazardly deploying a hose line and going straight for the "red" stuff. Put in plain English, suppressing a fire requires the following steps:

1. Find the fire (locate).
2. Assess the scene for victims (rescue).

3. Halt fire velocity and progression (contain).
4. Bottle up the fire into one generalized location/area (confine).
5. Mitigate or "go in for the kill" (extinguish).

Locate

The first objective of any fire scene or emergency situation is to locate the hazard. In the case of a structure fire, finding the fire may be as easy as seeing the glow and plume of smoke immediately as you leave the engine house bay (fig. 1–8). In other instances, it may be not so easy and may require a comprehensive size-up. Depending on the size and height of the building, personnel from several companies may be needed to conduct the size-up. Regardless of the difficulty, locating a fire should not be taken for granted. For instance, an engine company arrives on the scene and observes smoke coming from the front doorway and adjacent windows. The officer becomes a "moth to the flame" and orders an attack line through the front door without accurately conducting a thorough, 360-degree size-up. The fire, actually located in the basement, has been consuming the floor joists, and now the crew has fallen through the floor. Unfortunately, they located the fire; however, not in the manner they would have preferred.

Fig. 1–8. Finding the fire may be as easy as seeing the glow and plume of smoke as you leave the engine house bay. (Photo by Steven Heidbreder)

Many times, locating the fire requires experience, knowledge of building construction, understanding of smoke conditions, and preplanning. Often, it requires coordination of effort by multiple companies, sometimes from varied departments. Chapter 3 discusses size-up in depth, with key components for determining fire location.

Rescue

As stated in the fire service mission, life safety is the absolute, number one priority. Rescue must be made a primary objective if any of the following are true:

- Occupants are believed to be in a structure
- It is unknown whether there are occupants
- The probability of victims is significant

The caveat to initiating rescue procedures is the sustainability of life in the current fire conditions. For example, the engine company arrives on scene and finds intense heat and fire conditions in the living quarters of a structure that has already experienced, or will soon experience, a flashover. The possibility that a human could survive this environment is non-existent. At this point, rescue is not an objective in this location.

Rescue can be accomplished with many different strategies and tactics. These often require the coordination of multiple resources to perform simultaneous fireground functions. Meeting this objective, like most firefighting objectives, requires informed decision making based upon available resources and situational conditions.

Contain

When people envision firefighters responding to a fire, the first thing that comes to mind is that the main objective is to extinguish a fire. Ultimately, extinguishing the fire and mitigating the hazard is the desired outcome. However, before the fire can be extinguished, it should be controlled and stopped. Once again, there is a "moth to the flame" mentality in this case that should be avoided. The mission is to save and to protect property. Stopping the spread of the fire is the goal, not to extinguish what has already been destroyed and allow the fire to consume more.

Example: The engine company arrives on the scene of a well-involved two-story residential structure, with fire through the roof and burning through the second story on side B. The home is in a subdivision with small lots, tightly packed together. An exposure, another two-story structure, is 15 to 20 ft away and already off-gassing from the radiant heat. The officer has a decision to make and in this scenario only one is correct. Does the officer directly attack the fire in the involved structure or elect to protect the exposure by placing a line between the two houses? The answer is the latter choice. The officer should protect the exposure.

In this example, if the officer makes the right decision to protect the exposure, he/she has made a decision to contain the fire to the structure of origin. If the initial decision is not for containment, now there will be two structures with damage, possible heavy loss, and nothing gained. This is a common mistake for first due engine companies.

Every company officer, when approaching any type of emergency situation, should ask the following three questions:

1. What have I got? (now)
2. Where is it going? (future)
3. How am I going to stop it? (the plan)

Containing a fire is all about asking what is going on and predicting how it will react. Containing is stopping and halting further spread and fire progression.

Confine

Once the fire is contained to the structure of origin, it should be confined to a particular area. The terms *contain* and *confine* are similar in meaning, but there are differences. The biggest difference between the two definitions is halting (confining) and holding (containing). Therefore, confining can be considered the act of keeping the fire in a certain location, or bottling-up. Often, it can be considered redundant to separate the terms containment and confinement. Regardless, the importance is to have objectives that specify and prioritize stopping fire spread.

Extinguish

Finally, once fire progression has been stopped, our attention can be focused on complete extinguishment of the fire. Extinguishment for ordinary structure fires, which primarily consist of class A materials, entails the application of a sufficient amount of water. This amount is generally expressed in gallons per minute and dependent on a few variables.

Progressing from the Mission to Implementing Fireground Actions

This chapter explains why the fire service exists. Plain and simple, we have a duty to our community. Our services can range in delivery; however, our responsibilities are constant from region to region, nation to nation (fig. 1–9). The public places trust in us as an industry to provide a blanket of security and protection.

Fire does not recognize handicaps and does not discriminate. It burns at the same temperature, produces the same deadly by-products, and ravages property in any part of the country or the world. It doesn't care if a firefighter is getting a paycheck. It doesn't care if the response team is an army or a rollout of five people. The fact remains the same: there is a list of responsibilities that must be accomplished.

The fire service has, through years of experience, developed philosophies and generally accepted methods for attacking fires. These general concepts hold constant, regardless of the fire organization's capabilities. The objectives of rescue, locate, confine/contain, and extinguish are necessary and should be followed in a progression at every fire scene. The means of accomplishing these priorities (attack modes/methods) is also universal.

Fig. 1–9. The fire service can range in delivery; however, our responsibilities are constant from region to region, nation to nation. (Photo by Erin McGruder)

If the mission, objectives (priorities), and fireground actions (functions) are the same for any organization, why do we need a strategies and tactics text for a specific region, such as suburbia?

The Suburban Challenge Revealed

Accomplishing goals and objectives on the fire scene requires the following formula:

	What is to be done
+	*How the task is accomplished*
+	*Who is to do it*
=	*Fire service objectives*

Suburban operations can deliver services in a completely different manner than their larger urban counterparts, but the *what is to be done* stays constant in the equation.

Typically the physical resources, *what is needed to accomplish what is to be done*, and the human resources, *who is to do it*, are what vary widely between urban and suburban agencies. Therefore, tactics need to be adjusted to accommodate those differences.

Fire service organizations first need to conduct situational awareness assessments to realize what their capabilities are. These capabilities can be measured by successful completion of goals and objectives during fireground activities or training. Once again, this variable is highly dependent on available resources, such as equipment and staffing. Chapter 11 discusses staffing in greater depth. We are not just talking about numbers, but training and experience levels as well. Secondly, the organization must assess what it is up against and what handicaps it is facing in their delivery. Examples include response area characteristics.

Finally, once the organization has completed a thorough and complete self-assessment, these findings should be correlated with necessary fire service activities/functions (what is to be done) and a plan of attack should be developed. In the fire service, the plan for fireground procedures is documented in the form of *suggested operating guidelines* or *standard operating procedures*. Regardless of the terms used, they are synonymous and represent the same objectives.

As John Norman asserts, "Let circumstances dictate procedures." Chapter 11, "Tactical Variables," explains the factors and circumstances that differentiate suburban operations from urban. The three major variables discussed are

- Personnel/staffing
- Available apparatus/resources
- Response area characteristics

These factors greatly affect suburban firefighting operations, tactical selections, and functional assignments.

Chapter 2

Developing the Game Plan: Suggested Operating Guidelines

Plan Methodology

The biggest challenge on the fireground is consistently implementing tactics and strategies that are safe and successful. After discovering the capabilities and limitations of the fire agency's staffing, apparatus, and response areas, a response plan can be developed. This plan should be dynamic and adaptable to the dictated situation, yet have a basis to enforce a generally accepted guideline. Suggested operating guidelines (SOGs), if written properly, can fulfill these needs.

Plan development, in relation to structural firefighting, begins with the identified missions and goals of the fire department. The department leaders formulate the strategies to achieve the desired goals and mission. The leaders then identify the tactics necessary for strategic implementation and break the tactics down into the required actions, known as fireground functions. *SOGs are a blueprint, a set of directions, for achieving goals and objectives through consistent task assignment.*

Suggested Operating Guidelines Defined

Suggested operating guidelines are generally accepted procedures, adopted by the leadership of the organization, used to accomplish given objectives. The main focus of a SOG is to promote a consistent operational delivery that assures accountability and safety. Command staff members and company officers promote consistency by adherence to accepted guidelines and enforcement of fireground discipline. The adopted guideline should be clearly documented and accessible to all members. Accountability occurs through functional assignment. Each participating group or company detailed to an incident should be assigned a prioritized task and held accountable for the designated action. Functions are prioritized by their level of added value toward safety and achieving the objective. However, the number one priority and objective is always safety, safety of the victims involved and safety of responders.

The word "suggested" is an essential element to the guideline. Every command officer, company officer, and veteran firefighter knows that each fire or emergency incident is unique. Therefore tactics may have to slightly deviate from normal operating procedures. For example, it might be a general accepted operating guideline to defer the permanent water supply to the second arriving engine. However, if the second engine is delayed or the volume of fire dictates an immediate master stream, the first arriving engine officer may elect to deviate from the norm and forward lay a supply line upon arrival. Suggested states what the organization would like to see transpire on a regular basis; it also allows for a margin of decision making. In many instances, the situation will dictate tactical decision making.

In order for the SOG documentation to be clear and concise, it must possess essential elements. These elements can include a purpose section, assigned tasks, and general statements:

- Purpose section: Included in this section are the mission, vision, scope, and objectives. The organization explains the reason for the SOG and the vision of desired goals. Objectives, priorities, and driving factors should also be stated.

- Assigned tasks: This segment details individual company assignments/tasks.

- General statements: Included in this area are accepted operating procedures that pertain to all companies/groups involved. Examples of such procedures include Mayday procedures, communication information (operating frequencies/channels), particular truck placement (ladder gets the front of the structure), and any other pertinent information that is not conveyed in the assigned tasks section. Also included in this segment are general limitation statements. For example, "This SOG supports one attack line and one backup line."

The key to any well-written document is effective communication. Effective communication can be accurately measured by how well the receiver obtains and comprehends the intended information. For this to happen, the organization must capture the desired objectives and translate them into a format that is easily understood by the prospective operational staff members. In the next sections we thoroughly examine this process.

Beginning with Goals and Objectives

Chapter 1 established that the objectives of all fire departments within the nation are relatively similar. The fire service has sworn an oath to attempt to save lives, protect property, and to minimize negative impacts to the environment. These objectives are used to define the mission (goals) of the organization and are achieved with the strategies of location, rescue, containment, confinement, and extinguishment. The actions undertaken to achieve strategies are known as tactics.

> Example: Residential structure fire
>
> Goals: Safely provide services dictated by the mission of the organization, instill public confidence, and reduce property loss
>
> Objectives: Save lives and protect property
>
> Strategies: Locate, rescue occupants, contain, confine, and extinguish the fire
>
> Tactics: Search and rescue, coordinated fire attack

SOGs are designed to translate goals and objectives into to the simplest of fireground functions. These functions are the specific actions taken to complete the required tactics. The SOGs help in the organization of fireground functions and provide a system for their individual achievement (fig. 2–1).

Goal ➡ Objective ➡ Strategy ➡ Tactics ➡ Fire Ground Functions

Fig. 2–1. The relation of goals to fireground functions

Example: Residential structure fire

Goals: Provide services dictated by the mission of the organization, instill public confidence, reduce property loss, provide safety

Objectives: Save lives and protect property

Strategies: Locate, rescue occupants, contain, confine, and extinguish the fire

Tactics: Search and rescue, coordinated fire attack

Function: Forcible entry, deploy initial attack line, ladder deployment

The importance of breaking down this progression is to determine what exact actions are needed, then to determine exactly how they will be accomplished. Tactics take on a whole new meaning when functional assignments are handicapped by the opposing variables.

The difference between urban, suburban, and rural operations can be noted at the tactical/functional level. Therefore, suburban SOGs cannot be written in the same manner as the urban counterparts.

Prioritizing Function

Success on the fireground depends on maximizing the available resources. This maximization requires prioritization. Therefore, past experience must be reviewed to determine which functions take priority. For structural firefighting, past experience tells us no action taken on the fireground saves more lives than the proper size attack line, stretched to the correct location, and placed into service at the proper time. There are rare exceptions, which are discussed in chapter 9, where the attack line will be abandoned for straight rescue; however, the majority of time, the initial attack line is instrumental in the rescue function (fig. 2–2). Embracing this concept, the functions that support the attack line and contribute to the coordinated fire attack will have a higher priority, unless a rescue not involving the attack line is dictated.

Hey, let's get this one in place first.
I may need some help.

Fig. 2–2. Coordinated placement of the primary line may be critical to the success of your incident. (Illustration by Fire Medic Art)

The desired result during a structure fire is a coordinated fire attack. This term must be further defined and each element, or function, must be identified. A coordinated fire attack is the simultaneous action of particular fireground functions undertaken to promote safe operating procedures and maximize fire department resource potential to achieve the goals of rescue, containment, confinement, and extinguishment on a given fireground incident. Essential functions are those that promote victim removal/rescue, place the adequate size attack line in the correct position, and assure firefighter safety. The majority of time, these actions must be done concurrently. Operating procedures must be established to guarantee this occurs accurately and safely.

Many actions must transpire before, during, and after attack line placement to achieve operational success. These elements of the coordinated fire attack can also be designated as *facilitating* functions. In larger urban fire departments, these supporting roles are performed by ladder or truck companies. These companies typically do not resume any responsibility for stretching attack lines. Agencies that do not operate dedicated truck companies must develop systems to assign and prioritize these particular functions.

Before any prioritization can occur, all fireground functions must be identified. The list can be broken into four categories for simplification. The first category of fireground actions is *general functions*:

- Size-up
- Command
- Safety
- Rapid intervention team

The second category includes those functions typically assigned to *engine companies*:

- Primary attack line/stream deployment
- Water supply
- Backup line
- Additional attack lines
- Auxiliary systems (standpipe/sprinkler)

Truck company/facilitating functions make up the third category:

- Search and rescue (primary and secondary searches)
- Fire location
- Ventilation
- Forcible entry
- Utility control
- Ladders
- Overhaul
- Salvage

The final category is classified as non-traditional; however, necessary for both victim and firefighter safety. This category is the *emergency medical services*:

- Rehab
- Triage
- Treatment and transport

Before fireground functions can be further prioritized, they must be thoroughly analyzed and understood. The following section defines each function associated with fire suppression. Although these functions are common to urban, rural, and suburban firefighting in regard to structure fires, their applications in these settings differ.

Fireground Functions

Size-up

Size-up is the first fireground function that takes place on the fire scene. It is the ongoing collection of all possible signs, facts, conditions, and determinants which contribute to and affect fireground scenes and operations. Size-up begins during building and incident pre-planning and concludes at the termination of the event.

There are many elements of an emergency incident scene that are considered when formulating a thorough size-up. A popular mnemonic, first published by John Norman in his book *Fire Officer's Handbook of Tactics* (PennWell), is a foundation for remembering these essential fundamentals. This mnemonic is COAL WAS WEALTH:

- Construction type
- Occupancy
- Apparatus/personnel
- Life hazard
- Water supply
- Auxiliary appliances
- Street conditions
- Weather
- Exposures
- Area
- Location of fire
- Time of day
- Height

Many of these elements should be considered prior to the receipt of the alarm, during pre-incident planning. Physical conditions and characteristics of major target hazards and reoccurring incident regions should be noted and accessible to all members. These include building construction types, occupancy classes, life hazards, water supply limitation/capabilities, location and existence of auxiliary appliances, and street conditions.

Upon receipt of the alarm, company officers have the opportunity to receive initial information about the incident and coordinate this with the pre-incident plan. The dispatch should reveal apparatus responding, further locations, and time of day. Many computer aided dispatching programs can include additional information such as building layouts and special notations. These building preplans should include water supply locations, auxiliary systems, target hazards, and any other pertinent building features. Include special notations of items that could hinder operations, including street conditions or special permits. Extreme weather conditions can also be evaluated at alarm time; however, weather should be a consideration at the beginning of the shift period.

Once responding, companies should continue monitoring for possible clues that will improve their situational awareness. Company officers should listen to further radio traffic from dispatchers and earlier arriving apparatus. Hints of a working fire can include multiple calls to the dispatching agency, calls from neighboring addresses, police reports, and a visible column of smoke. Company officers should also be spotting hydrants, or note lack of hydrants.

Upon arrival, crews should observe various building characteristics and initial findings. These should include construction type, occupancy, existing exposure problems, area (size), location of the fire, smoke characteristics, and building height. The first arriving officer should attempt to locate the fire, determine how much of the structure is involved and associated life hazards, determine where the fire will spread, and develop an initial plan for mitigation. Clues that can determine occupancy are time of day, locked doors, and parked vehicles. A structure fire at 2:00 a.m. with locked doors and vehicles in the driveway should paint a different picture of occupancy as compared to a commercial structure fire at the same time of day with a vacant parking lot. Predicting fire spread, extent, and possible fire events (such as flashover, flame over, rapid fire spread, and backdraft) should be conducted by reading smoke. This is done by evaluating the color, volume, density, and velocity of the smoke.

Communication is an essential element in the size-up. All members conducting operations on the fireground should have an operational awareness of the situation. The first radio traffic from the initial arriving apparatus should be clear and concise. It should include apparatus identification, the structure type (including construction type, occupancy), structure size (height and estimated area), initial observations, initial actions and mode of operation (offensive, defensive, or transitional), and further orders for incoming apparatus.

Size-up is continuous. The incident and building should be closely monitored at all times, both internally and externally. Fire conditions can rapidly change during operations. Building integrity can also deteriorate from both the effects of heat and fire and firefighting operations. Member accountability should also be monitored for the entire duration of the incident.

Command

Command and control of an emergency incident scene should begin upon arrival at the scene by the first company or chief officer. All actions taken should follow a strict command structure that follows a traditional incident command system, mandated by the National Incident Management System (NIMS). The command function is typically transferred to a ranking chief officer upon his or her arrival at the incident. This individual has supreme authority over the incident.

Once the incident command position has been assumed, the command structure should be established. This structure should include the establishment of divisions and groups. Under the span of control, a single individual should have no more than five to seven reporting subordinates. Therefore, manpower permitting, individuals should be assigned to oversee and lead the divisions and groups.

Safety

Safety on the fire scene is every operating member's responsibility. Formally, the task of safety officer should be assigned to one person. This designee has the responsibility of monitoring operations and incident conditions (fig. 2–3). The safety officer has the authority to terminate any activity that is deemed hazardous.

The safety officer position, typically assigned to a later arriving battalion or division chief, can alternately be assigned to a capable company officer or experienced firefighter. This individual must be trained in the roles and responsibilities associated with the position, which is established under *NFPA 1521, Standard for Fire Department Safety Officer.*

The safety officer should begin by monitoring operations and determining that all operations are within safe operating parameters. He or she should also monitor all staff members for their physical condition, proper use of personnel protective equipment, and safe operating procedures. This position should also continuously monitor building and fire conditions and be in contact with the rapid intervention team through its designated officer.

Fig. 2–3. A safety officer looks on during fireground operations. (Photo by 5280 Fire)

RIT

A rapid intervention team (RIT) is a function designed to rapidly extricate firefighters in distress from hazardous environments. The key elements to its success are preparedness and operational readiness.

At least one RIT should be assigned at every incident, and the team should be properly staffed for the function of rescuing trapped firefighters. *NFPA 1500* sets a standard for minimal staffing in relation to RITs; however, each individual agency should determine its own staffing levels in accordance with its capabilities. Not all RITs should be the same and not all individual firefighters are trained to fulfill the role. All RIT members should be trained above standards for basic firefighting and be in compliance with *NFPA 1407, Standard for Training Fire Service Rapid Intervention Crews.*

RITs must be prepared and operationally ready. Preparation includes proper staffing and training. To be operationally ready RITs must have a situational awareness of the incident and structure and be in contact with the safety officer. They must also have a physical readiness, which means they should not be involved in activities directly related to operational activities. They should be reserved for activities that prepare the building for safe egress for firefighters, such as ladder placement, further forcible entry (including removing barricades such as iron window bars), and familiarizing the location of activities inside the structure.

Most importantly, the RIT should be prepared by assembling all essential tools for rapid intervention in an accessible location. The tool selection should be made with regard to the type of structure. For example, a crew assembling equipment for a commercial metal building should bring a K-12 saw with a metal cutting blade instead of a chainsaw designed for ventilating wood structures.

RIT tool suggestions include the following:

- Forcible entry tools (Halligan, axe, sledgehammer)
- Saws (chainsaw, K-12)
- Extra bottle or RIT pack (with full SCBA)
- Search rope
- Ladders
- Stokes basket
- Harness
- Hydraulic tools (dependent on construction type)
- Thermal imager

The building type and incident situation will dictate tactics. A single residential structure with minimal crew operating on the interior may only dictate the formation of one staffed RIT. However, a large commercial structure with numerous crews and a large operational area may require more than one RIT. Consider placing a RIT on every division for a large commercial structure fire.

Primary attack line/stream deployment

As stated previously, no action taken on the fireground saves more lives than the proper size attack line, stretched to the correct location, and placed into service at the proper time (fig. 2–4).

Fig. 2–4. Deployment of a primary line. (Photo by 5280 Fire)

Hence, the use of the primary attack line is the most critical function that takes place on the fireground. Its deployment is extremely time sensitive and failure can mean the difference between a successful operation and the creation of a parking lot. There are specific elements that go into ensuring a properly placed attack line:

- Effective gallons per minute
- Delivery method

- Nozzle selection
- Diameter
- Length
- Placement

Before an attack line is deployed, the initial arriving company officer must make an accurate size-up and decide which mode of operation will be pursued. The attack mode is either offensive, defensive, or transitional. This strategic decision strongly affects the choice of delivery method and placement. For an offensive attack, an attack line of either 2½ in. or 1¾ in. diameter will be utilized for interior maneuverability and mobility. If the mode is defensive, the initial weapon of water delivery may be a master stream in the form of an elevated stream or monitor. It may also dictate the immediate protection of exposures, which would give specific orders for the placement of the stream. If the mode is transitional, which begins as defensive and transitions into defensive possibly due to fire volume, a 2½ in. hand line may be deployed because of its potential for volume delivery and mobility.

The size of the attack line and nozzle selected will be determined by the gallons per minute required to achieve extinguishment. The fire officer must assess the area (length × width) and number of floors of the building consumed by fire and the materials involved. Keep in mind, however, that if there is an expected increase in reflex time (time from arrival to placing water on the fire) the officer should take fire spread into consideration when estimating the volume of fire to be extinguished. When considering materials involved, calculate commercial structure contents as producing more BTUs than residential, requiring more gallons per minute.

The length of the attack line must be estimated for successful water delivery to the fire. To accomplish this task, the company officer must make the following observations:

- Distance from the apparatus to the entry
- The area of the fire floor (length × width)
- Elevation (floor of fire location)
- Distance from entry to stairwell (if the fire is on a different grade than the entrance).

For each floor, below or above grade, one section of hoseline (or 50 ft) should be added to the formula when estimating hose stretches. All factors are added together to get a total length of stretch. The company officer must practice estimates and become familiar with distances by using common references.

For example, the width of the engine house parking lot may be close to 200 feet. The officer can use this as a mental yardstick for practical application.

Proper attack line placement is essential. The key to success is reducing flex time. The placement that puts water on the seat of the fire in the quickest manner, from time of arrival, is the optimal tactic. Crews are not only attempting to achieve containment, confinement, and extinguishment, but they are also attempting to save/rescue victims while staying safe in the process. This can be achieved by placing a hose stream between the fire and those still in the building (victims and firefighters). Another factor that should be considered is gaining entry through the front door on residential structures. There are two reasons for this:

- A large majority of unconscious occupants are found in the path of common egress
- Protection of the stairwell (stairwells are commonly found adjacent to the front door).

As discussed earlier, mode will be a determining factor in stream placement. An offensive operation will most likely require an interior placed attack line, while defensive and transitional operations will dictate an initial exterior placement.

Water supply

NFPA 1710 states that there shall be an uninterrupted water supply of a minimal of 400 gallons per minute for at least 30 minutes. Initial companies can accomplish this task by attaching an adequate sized supply line(s) to a hydrant(s) or set up a temporary water supply system, such as a tanker-shuttle operation for areas without sufficient public water mains.

For practical application, initial companies need to exercise *down-board* thinking when considering water supply. The first critical action that should take place by the first arriving officer is to estimate the gallons per minute necessary to extinguish the given fire. After selecting the attack line of choice, the officer must then decide how that attack line will sustain operation given the water supply. In many cases, the initial officer will elect to defer the permanent water supply responsibility to the second due engine and initially operate off booster tank water. If this is the case and this officer is flowing 150 gpm with the initial attack line from an apparatus with a 750 gallon booster tank, the second due engine has 5 minutes to establish a water supply without interruption. A trick most good engine operators use is an immediate supply line from apparatus to apparatus, giving the second engine's booster tank water to the attack line apparatus, while simultaneously securing a water source. This buys an additional 5 minutes if the second engine has the same booster tank capacity.

When securing the water source from a hydrant, there are three methods:

- Forward lay
- Reverse lay
- Combination

The forward lay has become the standard for many fire departments. For this operation, an engine will stop at the hydrant before arriving at the fire structure and lay the supply line toward the fire. The reverse lay accomplishes the opposite. The engine arrives at the fire, pulls off all necessary attack line, and then the engine proceeds to the hydrant. Combination lays are used by two engines, one using a reverse lay and one using a forward lay, to meet somewhere in the middle to cover long supply distances.

Consider this example: A 500 gallon booster tank engine arrives at a structure with heavy fire showing. The officer decides to pull a 2½ in. attack line and flow 500 gpm. This officer now has 2 minutes of flow time. The best option for this apparatus is to secure its own water source. The officer should either stop at the hydrant and forward lay a supply line or pull off enough sections of 2½ in. at the front of the structure and reverse out to the nearest hydrant. The key to the reverse out is having enough of a static load of 2½ in. hose.

Another consideration is having enough capacity and pressure. Officers must have the ability to recognize the gpm that each hydrant is capable of flowing. In most areas, this is identifiable by a bonnet color system. If the structure in question has a large area involved, or large fire load, more than likely multiple hydrants and mains will be needed. If more pressure is needed, the officer should considering placing a supply pumper attached to a hydrant to boost the pressure.

Not all regions or areas are fortunate to have publicly supplied water sources. It is especially important to preplan and develop special guidelines and SOGs for structural firefighting in these areas (fig. 2–5). Uninterrupted water supply, regardless of lack of hydrants, is still essential. Special tactics for handling suburban water supplies are detailed in chapter 6.

Fig. 2–5. Not all regions or areas are fortunate to have publicly supplied water sources. (Photo by Erin McGruder)

Backup line deployment

Backup lines are essential for firefighter safety and readiness. They serve the contingency plan if initial tactics are not adequately achieving strategic goals. It is important to remember that *NFPA 1710* dictates crews should be able to apply 300 gpm through at least two hand lines, regardless of the size of fire. The backup line should be equal to or greater (in diameter and capacity) to the initial attack line (fig. 2–6).

Fig. 2–6. Firefighters deploy an additional long lay. (Photo by 5280 Fire)

Backup lines should be in place in case situations worsen or tactics fail. They should be in a position to support or protect the attack line. An example of this protection includes the use of backup lines to defend and control stairwells. If backup lines are placed into operation as attack lines, considerations should be made to staff another line. Also, a good operating policy is to mandate additional lines from a different engine company, separate from the initial attack line engine. This action safeguards against operational pump failure.

Additional attack lines

Once the initial attack line has been placed, conditions may dictate that crews initiate the deployment of additional attack lines or streams. These conditions may include

- Additional gallons per minute needed for extinguishment
- Exposure protection
- Multiple floors/grade of fire
- Rescue situations (including firefighters)

Auxiliary systems

When responding to commercial structures, crews should consider the possibility that the building has an auxiliary fire system. These systems can be in the form of standpipes, sprinkler systems, or a combination of both.

Fire crews should preplan their respective response areas and pinpoint all auxiliary fire systems present in commercial buildings. During fire operations, these systems will need augmentation by fire apparatus. Engine companies will need to attach supply lines and augment the water supply from nearby hydrants. In some instances, fire department personnel will be needed to operate interior system fire pumps.

Suppression operations initiated from standpipe systems require many additional considerations. This topic is thoroughly covered in chapter 7, "Special Considerations: Commercial Structures."

Search and rescue

Fire agencies should not forget their primary mission: to save human lives. Search and rescue, in relation to structural firefighting, entails searching for victims who are either overcome or trapped in a fire involved structure and removing them from hazardous environments. There are two phases of searching:

- The primary search
- The secondary search

The primary search is a quick search of probable areas where victims might be found. It is less methodical and *fast*. The key is to find the victims closest to harm's way and to remove them before they perish. To accomplish this task, the company officer must acquire as many facts as possible before making entry. These facts will help determine the primary search:

- Time of day
- Reports from bystanders
- Locked doors
- Previous knowledge of the building layout
- Occupancy type
- Parked vehicles
- The number of possible victims

The primary search occurs during the fire involvement and before the fire is under control. The officer must make a decision to search probable areas where victims will be located and from the seat of the fire back toward less affected areas.

Secondary searches are methodical and more time consuming. Crews should attempt to search the entire structure thoroughly. Primary and secondary searches should not be conducted by the same crew. This is done in order to get a second set of eyes and a different perspective. Secondary searches usually occur after the fire has been placed under control and the heat and smoke condition has been removed.

Search and rescue can be accomplished through many different tactics. There are some common search and rescue tactics:

- Truck company searches in front of the hoseline
- Searching from the nozzle back
- Searching above the fire (the floor above)
- Vent, enter, and search
- Ladder rescue

Vent, enter, and search (VES) is the act of venting a single room (the majority of time from a ground ladder), entering the room (without a hand line), confining the room by shutting the room's door and/or windows, and conducting a quick primary search of that particular room, and exiting the room back through the same point of egress.

The tactics chosen are usually dependent on available, initial arriving firefighters, proficiency and knowledge of firefighters, and the situation. If your agency does not staff truck companies with engine companies, initial searching without an attack line is not advocated. It is also not a wise decision to conduct searches above a fire, or VES procedures with inexperienced or non-proficient firefighters. Conducting an initial ladder rescue, with minimal resources, multiple victims from multiple windows, and a progressing fire without an attack line deployment is also a bad decision.

Fire location

Locating fire is often taken for granted and underestimated. The location of the fire is not always detectable upon initial size-up. Deployment of resources, such as initial attack line placement, without knowledge of the fire location can be disastrous.

The initial company arrives on the scene and attempts to make an accurate size-up. The officer attempts to establish answers to three questions:

- What have I got? (Determining what the situation is.)
- Where is it going? (Determining spread and future fire events.)
- How do I stop it? (Determining what strategies to implement to mitigate the situation.)

If the officer is lucky, he or she might have visible fire showing that pinpoints the exact fire location. However, this is not always the case. The situation may be that only smoke is visible from the exterior (fig. 2–7). In this case, the officer should be able to decode the characteristics of the smoke. Fire will generally be located in the site of the darker, higher velocity smoke. Because velocity will develop due to heat, high velocity indicates a close proximity to the fire. The darkness of the smoke indicates heat and carbon particles in the smoke. As the smoke travels any distance away from the fire, the potential increases for the carbon particles to be scrubbed out, reducing the black color. Regardless of visible fire or smoke qualities, a 360° size-up still must be conducted.

There are some cases where fire or smoke may not be visible. The fire may be in a larger complex away from the main entrance or in a multistory building, where the fire room is not visible from the approach street. Investigation may require a little more effort and time in this situation. It might require more than one crew. In the case of the large complexes and multistory buildings, it is always a good tactic to have two arriving apparatuses approach from opposite directions to get a good full view of the structure. The alarm panel and enunciator panel is also a critical tool in this scenario. If nothing is visible from the exterior, the initial arriving officer should immediately check the alarm or enunciator panel for problem locations.

Fig. 2–7. In the instance only smoke is visible during the 360° size-up, the officer should be able to decode the characteristics of the smoke. (Photo by Steven Heidbreder)

Not all fire location happens from the "front seat" of the apparatus. After the officer conducts the 360° of the building and decides to proceed through the front door, the officer and crew should not hastily rush through the door. Once the door has been opened, the crew should observe the smoke. If the smoke rises to the midway point of the door, it is a good indication that the fire is located on the same level as entry. If the smoke rises to the ceiling and dissipates, there is a strong hint that the fire is above grade. Finally, if the door opens and the smoke does not dissipate and hangs at the floor level, be very careful. This is a sign that the fire is possibly below grade.

There should be a crew assisting with fire location during interior attack operations. A fire that escalates past the room and contents has the potential to get into the walls, ceilings, and void spaces and has the potential to get behind and cut off an attack line. Fire location crews should have the appropriate tools to check these areas as the hose team proceeds. The same principle can be applied by crews attacking a fire on multiple levels. Crews should not proceed to another level until the fire is contained below them. Therefore, interior crews must be constantly pulling ceiling and walls to ensure safety of the advancing hose teams. This is a critical *facilitating* function that must not be overlooked.

Ventilation

The International Fire Service Training Association (IFSTA) in chapter 11 of *Essentials of Fire Fighting and Fire Department Operations*, 5th edition, defines ventilation as the systematic removal of heat and smoke and replacing with fresh air. Ventilation is performed for three primary strategies:

- Ventilation for life
- Ventilation for property
- Ventilation for firefighting (suppression activities)

For practical application, ventilation can be accomplished vertically or horizontally. Tactically, it can be performed naturally (using natural wind currents) or forced (using positive or negative pressure, or hydraulically).

Ventilation can save lives and facilitate fire suppression activities. If done incorrectly, however, it can spread fire and be dangerous for interior crews and trapped victims. Ventilation is extremely time sensitive. If it is initiated before attack lines are in place, the fire will spread rapidly and crews will lose any chance of initial containment. The introduction of fresh air and oxygen will fuel the fire and push the fire in the direction of wind current flow, from areas of high pressure to low pressure. Ventilation is a fireground tactic that requires precise coordination with attack crews.

Crews must decide how ventilation will be conducted, considering the following factors:

- Purpose (vent for life, property, or firefighting operations)
- Location of fire
- Location of victims
- Structural conditions
- Exposures
- Weather (wind conditions)
- Available staffing

When venting for life, crews must consider possible victim location and attempt to get fresh air to them as soon as possible, without spreading fire toward them. Crews should consider horizontal ventilation due to the fact that this can be accomplished quickly and it requires less staffing. Also, consider possible VES operations if crews are proficient and trained for this activity. If considering positive pressure ventilation, crews should be extremely cautious and even reconsider. Positive pressure ventilation will push and spread fire significantly more than natural techniques.

Firefighters can use ventilation as a tactic that assists in property conservation. Venting for property means that crews are attempting to stop the spread of fire by removing heat and smoke or redirecting the heat and smoke away from unburned areas. The type of ventilation used will be largely dependent on fire location, building conditions, exposures, and time sensitivity. If the fire is located in areas close to the roofline, such as the top floor or attic, consider vertical ventilation. To conduct this tactic, crews must have enough time and staffing, as well as a structurally sound roof, to affect operations successfully. Trenching is a vertical technique used to stop fire spread in large, long structures. This subject is addressed in depth in chapter 7. If the fire is involved in a structure with close exposures, horizontal ventilation may be contraindicated.

Ventilation tactics are used in facilitating fire suppression activities. Venting for firefighting is an essential element in the *coordinated* fire attack. Removing smoke and heat will assist suppression efforts:

- By increasing visibility
- By redirecting heat away from the attack crews
- By reducing the potential for steam burns for interior crews

The key factor to this facilitation tactic is coordination and communication. Ventilation should not occur until all attack lines are in place and ready for operation. Communication between the attack line officer and the ventilation officer is vital.

Forcible entry

Nothing can occur on the fireground until access to the structure is obtained. Forcible entry is the act of gaining access where entry is either locked, blocked, or nonexistent. For safety purposes, forcible egress has also been added to the function of forcible entry.

Forcible entry success depends on crews knowing the construction features and mechanisms that keep entry to structures secured. Secondly, crews must have the knowledge of techniques used to force entry and the correct tools to perform these techniques. Not only must they have the right tools for the job, they must have them assigned, present, and ready on the fireground.

Forcible entry can take the form of

- Forcing doors and windows
- Breaching walls
- Cutting/breaching fences
- Breaking locks/padlocks

Initial forcible entry techniques, in relation to structure fires, include

- Conventional forcible entry
- Breaking glass
- Using through-the-lock and other lock breaking tools
- Using powered tools

Utilities

During suppression activities, crews must be assigned the task of utility control for safety considerations. Electricity, gas, and water must be shut off to the structure to enhance operations. For buildings that contain elevators, elevators must also be controlled. This operation will require one assigned firefighter to oversee the function.

Ladders

Placing ladders at a structure fire is an essential function that must be completed in order to facilitate the following functions:

- Extinguishment/suppression
- Rescue
- Ventilation
- Safety

Crews use ground ladders on the majority of fireground operations. Gaining access to floors above grade or to roofs has to be accomplished expeditiously. Ladders must be placed on apparatus so that they can be accessed within a moment's notice and all crewmembers must be proficient with their deployment and use. When rescues requiring ladder deployment are found upon arrival, crews only have seconds to react.

Even though ladders are used at every fire scene, they are often under-deployed. Ladders are usually raised to accomplish a function. But the one function that is overlooked is safety. Any time crews are working above grade, where areas are accessible by ladders, each side of the building should have a ladder raised for an emergency egress. This can be accomplished with either ground ladders or aerial devices.

Overhaul

Overhaul can be defined as the search for hidden fires. Many texts, such as *Essentials of Fire Fighting and Fire Department Operations* (IFSTA), suggest that the purpose is to find hidden sparks to prevent rekindle or to find areas of origin for fire investigation. Often, overhaul is considered a discipline of *loss control*. Searching for hidden fires entails more than preventing rekindles and searching for areas of fire origin, though. This function coincides with locating the fire.

Overhaul is actually conducted during two phases: presuppression and post suppression. During suppression activities, crews should be pulling ceilings and walls to find any hidden fires in void spaces. As stated earlier, if fire were to break out of these void spaces, advancing hose crews could become trapped. Overhaul also occurs after the suppression activities to ensure all hidden fires and void spaces are completely extinguished. After the fire is under control, crews can also assist investigators find areas of fire origin by opening up walls, floors, and ceilings.

Salvage

Salvage is an element of loss control designed to reduce damage to property from both fire events and fire suppression activities.

Fire department personnel should not lose focus of the primary objectives of fire suppression. That objective is to support the mission of protecting and conserving property. Often, this focus is lost and fire suppression activities, such as hoseline application, forcible entry, and overhaul, causes more damage than the actual fire. If possible, crews should immediately consider property conservation and take actions, such as salvage cover deployment, as early as possible. Realistically, however, suppression activities take precedence due to safety factors.

Salvage procedures should be conducted at every occupied structure fire. Loss control ensures public confidence in the fire agency and adds value to operations.

The following three functions are activities related to emergency medical services (EMS). Regardless of whether your organization operates fire-based EMS systems or contracts the service through a private agency, these functions must be considered at structure fires.

Rehab

Considerations should be made at every structure fire to monitor the health and well-being of all firefighters. Also, provisions should be in place to rehydrate all physically exerted operational members and allow time for recuperation.

Rehab on a fire scene should be more than simply getting a 5-minute break or changing an air bottle. Experience and statistics have shown that the largest category for fireground fatality is heart related. The exertions of fire suppression activities can be overwhelming. All firefighters should be monitored by at least having their heart rate, blood pressure, and respiration rate assessed. Oxygen saturation and heart monitors should also be available to further assess those firefighters showing signs of excessive stress and exertion.

Triage

If a structure fire occurs with the possibility of multiple victims or mass casualties, crews must conduct triage activities. Triage is the act of assessing victims and determining a priority for treatment and transport. This function can be staffed by firefighters trained as EMTs or EMS personnel.

Treatment and transport

Ambulances must be initially dispatched on all structure fires for the possible need to transport occupant victims or injured firefighters. EMS crews must be available to immediately treat all injuries and to assist the transport to the appropriate facilities.

After understanding fireground objectives, what tactics are necessary to accomplish these objectives, and defining each necessary fireground function, we can identify a general order, or priority, for these tactics. Each function is necessary; however, some more time sensitive than others.

The Fireground Priority List

Once fireground objectives, the tactics necessary to accomplish these objectives, and each necessary fireground function are understood, we can identify a general order or priority for these tactics. Each function is necessary, however some are more time sensitive than others. The fireground priority list begins with the strategies and functions dictated by the fireground objectives:

- To save lives
- To conserve property
- To limit the harmful impact to the environment

In regard to structural firefighting, two general strategies take precedence:

- Rescue victims
- Initial line placement (dependent on mode)

These two strategies are so imperative and time sensitive that they must be initiated by the first arriving apparatus or engine. The only time that initial line placement would not be attempted by the first arriving apparatus is if that apparatus is a dedicated truck company or rescue is immediately initiated without line placement. Chapter 4 contains a thorough review of the decision-making process between rescue and line placement when faced with limited resources. Therefore, the following list begins with the functions that are conducted to initiate the strategies.

Functional priorities for suppression apparatus

Functional priority 1: awareness

- Perform the size-up.

Functional priority 2: command

- Take command.
- Determine the attack mode.

Taking command in this context does not signify permanent incident command. This denotes taking immediate responsibility for the fireground and making initial decisions (fig. 2–8). It is also not indicated that this initial arriving officer will remain on the exterior during initial offensive interior attack operations. It is far more dangerous to wait outside for incoming crews to assume interior duties or to abandon the crew in order to assume exterior command.

Functional priority 3: rescue

- Initiate initial search and rescue.
- Force entry.
- Place ladders.
- Place lines.

For many suburban applications where only one apparatus arrives on the scene initially (no truck companies), the majority of the time rescue will be achieved only through initial line placement. Rescue can only be accomplished from the exterior in this situation due to lack of resources and the protection of a hoseline.

Fig. 2–8. Initiating command means taking immediate responsibility for the fireground and decision making. (Photo by Box Alarm Productions Fire Ground Photography)

Functional priority 4: place the initial attack line

Functional priority 5: facilitate the attack line

- Secure water supply.
- Complete forced entry.
- Assist with line placement.
- Ventilate.
- Assist with fire location (opening walls and ceilings).
- Assist with search and rescue.
- Place additional ladders.

Functional priority 6: deploy additional line

- Place backup line.
- Place additional attack lines.

Functional priority 7: facilitate additional attack lines (*if applicable*)

Functional priority 8: complete all incomplete support (truck) functions

- Complete vertical ventilation (*if applicable and incomplete*).
- Assist with ladders.
- Secure utilities.
- Perform post-overhaul.
- Effect salvage.

Functional priority 9: RIT

Placing RIT as the last priority function is controversial. The importance should not be diminished, however. The justification is that firefighter safety should be a proactive action on the fireground. Good tactics and strategies, such as stopping fire spread or removing the hazard, are as much as a protective measure as RIT. *RIT is absolutely necessary*; however, it is a parachute for those firefighters in absolute desperation. Fix the plane before it goes down. It is far safer to minimize the Mayday first.

Functional priority list for command (chief) officers

Functional priority 1: command

Functional priority 2: division or branch assignments

Functional priority 3: safety

Functional priority list for EMS

Functional priority 1: triage

Functional priority 2: treat and transport

Functional priority 3: rehab

Rehab is prioritized last in the EMS priority list; however, is an essential function on structure fires. If the initial assigned EMS unit becomes involved in triage, treatment, or transport, it is necessary to request an additional unit.

Pairing Function with Assigned Apparatus

Now that functions have been prioritized in relation to strategies and objectives, they must be appropriately assigned. Policy makers must establish the desired amount of apparatus dispatched on the initial alarm and attempt to maximize resources. Most importantly, they must also realize the potential and limitations of each assignment. Many suburban fire agencies rely on mutual aid agreements in order to satisfy resource needs. Others may not have this luxury and depend on limited staffing and resources to accomplish the same demands.

Before assignments can be made, leaders must answer the following questions:

- Is staffing on a single resource enough to accomplish a single function?
- How many functions can a single resource accomplish safely and in a timely manner?
- Is your staffing educated and sufficiently experienced for every assignment on the fireground?
- How many attack lines can your initial alarm support?
- Do you have specific apparatus for specific functions (e.g., dedicated truck companies)?
- How many resources does it take to effectively handle all fireground functions for my agency?

To begin, consider the average urban response and attempt to translate that to a suburban application. Many urban fire departments that operate with dedicated truck companies may initially dispatch these resources:

- Three engines
- Two trucks (or ladders)
- One chief officer

Upon verification of a working fire, the urban department may upgrade with additional apparatus, including a heavy rescue company. The initial response may also include the heavy rescue. The fire department knows exactly who is going to perform engine work (deploying hoselines) and who will be performing truck work (facilitating assignments).

Now compare this to a similar suburban response that does not operate dedicated truck companies. This dispatch might look like this:

- Five apparatus (all of which have pumping capacities)
    - Engines/pumpers
    - Rescue/engines
    - Aerial apparatus (quints)
- One chief officer
- One ambulance

The functional assignment isn't as cut and dried in this situation. Functions must be assigned on arrival order and not by resource type.

To simplify functional assignments, suburban fire departments need to adopt a basic operating principle that is flexible yet assures the timely and safe completion of all prioritized fireground functions. One such system is *one-plus-one* structure. The basis of this method is there should be one *facilitating* apparatus assigned for every engine function or line deployment. Therefore, all functions have been accounted for with built-in flexibility. In the previous example, two apparatus will be assigned line placement, two companies will share the responsibility of facilitation (all support functions), and the fifth will assume RIT. If all functions are not satisfied, the incident commander should upgrade the alarm with more resources.

One Plus One

The basis of the "one plus one" method is that there should be one facilitating apparatus assigned for every engine function or line deployment.

Fig. 2–9. The one-plus-one concept can be leveraged to increase fireground safety. (Illustration by Fire Medic Art)

The underlying purpose of the one-plus-one method is safety. Hoseline crews should concentrate on the task of situational awareness and proper placement. When individuals are given too many tasks, they become overburdened and are unable to be attentive to detail. Unsafe tactics are conducted. The one-plus-one method assures accountability and a buddy system for interior crews. Hoseline crews should not conduct operations on the interior of the structure solo. They should have facilitating crews to assist with placement and to be available in case these crews face peril.

Suburban SOG example

First arriving apparatus (engine functions)

- Size-up
- Rescue
- Command
- Forcible entry
- Initial hoseline/stream deployment
- Consider water supply

Second arriving apparatus (facilitating/truck functions)

- Water supply
- Assist placing line (corners/doors)
- Search and rescue (primary search)
- Ventilation
- Complete forcible entry
- Fire location (pre-overhaul)
- Ground ladders

Third arriving apparatus (engine functions)

- Additional line deployment
    - Backup line
    - Additional attack line*

      *If this line goes into operation as an attack line, additional apparatus will need to be added to the initial alarm to staff a backup line. This five apparatus SOG will not support a third hoseline deployment.

Fourth arriving apparatus (facilitating/truck functions)

- Facilitate second attack line (if applicable)
- Complete all supporting (facilitating/truck) functions
    - Vertical ventilation (if applicable)
    - Ground ladder placement
    - Additional forcible entry
    - Utilities

- Salvage
- Overhaul

First arriving EMS unit

- Triage
- Treatment and transport of victims
- Rehab

Functions are not dictated by resource type; however, rules should be developed in the general statements section of the SOG that stipulate truck placement. For example, engines should still give the front of the building to aerial devices. These apparatuses may not operate as truck companies, but the vehicle still has the responsibility of aerial placement.

The key element in the suburban SOG decision making process is communication. Initial arriving crews must communicate needs and deficiencies to incoming resources so that they can decide what priority functions must be completed next. Critical communication includes items like the following:

- **Water supply.** Has water supply been established?
- **Mode of operation.** What initial attack mode has been selected?
- **Hoseline placement.** Has the hoseline made successful progress to the seat of the fire or does it need assistance? This is a critical factor. No additional hoselines should be initiated until this line is established. Only functions that assist and support the deploying of this hoseline should be initiated. *First line first.*
- **Ventilation.** Is immediate ventilation needed and what type?
- **Incomplete facilitating functions.** Which functions still need to be completed?
- **Additional attack lines.** Do additional attack lines need to be deployed?

The Specific Game Plan

SOGs should be created to support the needs of specific fire agencies. The organization should be able to identify the critical factors that affect the delivery of services in regard to structure fires. Agencies should be able to pair resources with the identified functions and develop a general game plan. If the agency has conducted a thorough and accurate operational awareness assessment, these functions can be assigned with the assistance of a SOG worksheet (fig. 2–10).

After the SOG has been developed, it is the responsibility of the organization to successfully implement it. The next chapters discuss procedures and implementation plans in detail.

Suggested Operating Guidelines for Structure Fires Worksheet

Fireground Functions

General functions
- ☐ Size-up
- ☒ Command
- ☐ Safety

Engine functions
- ☐ Primary attack line
- ☒ Water supply
- ☐ Backup line
- ☐ Secondary lines

Auxiliary (truck) functions
- ☐ Search and rescue
- ☐ Ventilation
- ☐ Locating fire
- ☐ Forcible entry
- ☐ Utilities
- ☐ Assisting w/hoseline placement
- ☐ Ladders
- ☐ Salvage and overhaul

RIT

EMS
- ☐ Rehab
- ☐ Transport

Assigned Apparatus

First-arriving apparatus WATER SUPPLY

Second-arriving apparatus ____

Third-arriving apparatus ____

Fourth-arriving apparatus ____

Fifth-arriving apparatus ____

First-arriving ALS unit ____

First-arriving chief officer COMMAND

Considerations LADDER TRUCK GETS THE FRONT OF THE STRUCTURE
SECONDARY LINES COME FROM DIFFERENT PUMPERS/QUINTS

Fig. 2–10. Example suggested operating guidelines worksheet

Part II

Application of Suburban Fire Tactics

Concepts in the fire service fail unless reinforced through effective fireground application. Fire service instructors, company officers, chief officers, and policy makers must be able to realize their agencies' capabilities and limitations and directly apply these to suggested operating procedures. The fire service, regardless of demographics, has always been posed with situational challenges. A fireground or any emergency can be directly influenced by the initial impact actions of the responding companies. This text applies the concepts and methodologies of customizing suggested operating procedures, for specific agency use, for modern day practical firefighting.

Part II begins with the basics of initial fireground functions—the size-up; prioritizes initial fireground actions by illustrating the key roles of the first and second arriving apparatus responsibilities; and then progresses toward more advanced functions such as technical rescue, search and rescue, and rapid intervention teams. It also includes concepts such as

- Primary decision making
- The facilitator
- Suburban water supply
- Commercial structures

This section illustrates and reinforces the need for suburban consideration and customized suggested operating procedures.

Chapter 3

Size-Up

Size-Up Overview

Size-up is the first action that must take place on the fireground or any other emergency incident scene. Company officers must gain a situational awareness of exactly what they are up against (fig. 3–1). This is a detailed process that entails many variables. No situation or incident is exactly the same, and many factors can influence emergency response events. This holds true for urban and suburban operations alike, and should never be taken lightly. Suburbia can present many surprises and challenges for company officers as well as emergency responders. The key is to gather all the pertinent information necessary to anticipate or predict events that will occur during the incident. Once this has been accomplished, tactics and actions can be conducted to effect safe operations for incident mitigation.

Size-up is the ongoing process of gathering all information that directly impacts the conditions and circumstances of a particular incident. Technically, size-up begins at the receipt of the alarm, even before arrival. However, let's take it a step further and make the comment that it actually should start much earlier than the receipt of the alarm. It should begin with district preplanning and response area familiarization. Size-up does not conclude until the incident has been completely mitigated and considered completed. It continues for the entire duration of the event because the event has the possibility of changing. In fact, it should change. The intent of the tactics and strategies that are being conducted is to positively change or alter the incident. It is important to continually evaluate the circumstances and conditions to make sure the tactics are effective.

Fig. 3–1. In absence of the visible fire, a company officer should be able to identify that the fire is located closer to the turbulent black smoke than the white laminar smoke. (Photo by Michael Thiemann)

What happens when size-ups are not conducted or are completed ineffectively? The entire incident has the possibility for failure. Take a look at this example:

> *The engine company arrives to find light smoke showing from the front of a perceived one-story residential structure. The company officer, in the excitement of a working structure fire, orders a line stretched before he jumps off the rig. The officer skips the 360° walk around of the structure, and the line enters the front door. Unfortunately, this structure is an atrium ranch, with a fire originating in a large basement area with rapid spread to the open first floor.*

The officer evaluated the fact that smoke was visible from the front of the structure and made the conclusion that: 1) there was a working fire in the structure and, 2) the answer was an immediate line through the front door. The situation gets worse, however. He unfortunately missed a few more circumstances that would have assisted with situational awareness and initial tactical decision making.

The missed circumstances include

- Building construction (this is an atrium ranch construction and not a typical one-story)
- The location of the fire
- The extent of the fire
- The stage of the fire
- Life hazards in the structure

Because of these missed conditions, many bad events can befall this initial arriving crew. If they are lucky, the best situation is the crew is able to endure the tremendous heat from the basement and reach the seat of the fire from their blind, difficult stretch into the basement. Because of the lack of situational awareness, much worse events can occur:

- The fire grows exponentially out of control because the crew is unable to reach the seat of the fire and the fire is in a heavy growth phase.
- The crew falls through the floor and into the basement because the fire has been rapidly consuming structural floor members and they were unaware that the fire was below them.

Implementing tactics and strategies at working structure fires without first collecting accurate circumstantial information is like driving blindfolded! Safety is compromised and fireground functions can prove ineffective. This is especially true in suburbia where variables (personnel, available apparatus, and response area characteristics) impacting tactical delivery are already creating challenges for operations.

Preplanning

Preplanning can be argued as the first step in the size-up process for any incident. There is no better time to gather information about an incident than before it occurs. All fire service agencies and personnel should participate in preplanning.

What exactly is preplanning? Is it a formal process? Preplanning can be both formal and informal. The formal process entails the documentation of critical hazards and structures within response areas.

Formal (documented) preplanning

Many fire agencies keep a formal documentation system of most, if not all, commercial buildings within their response areas. More advanced systems are computer based and can be maintained in a database. Today, many fire companies can have this information readily accessible at their fingertips while responding on alarms through onboard computers known as mobile data terminals (MDTs). Quick access to this vital information assists with tactical decision making and firefighter safety.

Information commonly contained in formal building preplanning includes the following:

- Type of construction
- Occupancy type and life hazards
- Building dimensions and layout
- Roof structure
- Exit/egress locations
- Stairwell access and location
- Fire department control panel locations
- Sprinkler system locations (including areas that are sprinkled, control room, main riser, and the exterior fire department location)
- Standpipe locations
- Water control valve locations
- Utility control locations
- Special building systems (elevators and control rooms, ventilation systems)
- Hazardous material locations
- Any target hazard
- Areas or rooms of special interest (nurse's offices, first aid locations, etc.)
- Fire flow calculations
- Available hydrant location and available water supply (main size, pressure, gpm)
- Exposures
- Access and street descriptions

Preplanning does not have to be limited to just building characteristics. Some agencies also choose to include standard tactical decision making to anticipated response locations. For example, the plan may call for specialized apparatus and may detail extra equipment to higher target hazards. It may also suggest equipment placement. An example of this is a building with limited access. Even though a good company officer should be aware of these conditions within his or her response area, the preplan can serve as a reminder or give information to incoming officers who are not familiar with the direct response area. Buildings that are not structurally sound and far too dangerous to enter under fire conditions should also have such preplans with warnings attached. These plans should include tactical action plans for defensive operations and exposure control.

There should be a word of caution, however, when making preplanned tactical action plans. Keep in mind that variables change and that plans should not include constants. For example, if your tactical action plan revolves around the placement of your only 100 ft tower ladder for above-grade access to trapped occupants and that apparatus is out of service or unavailable, what will you do? That plan is worthless and could delay or complicate operations. Be extremely careful with absolutes. No fire is the same and all conditions can change moment to moment. Remember, operating guidelines are developed with variables in mind and adaptation to change. *Preferred* and *suggested* are critical terms that should not be eliminated from the methodology of tactical planning.

Informal preplanning

Informal, or undocumented, preplanning is just as important as the formal documented version. This type of planning is more commonly known as district or area familiarization. Fire companies should be routinely making observations about their response areas. Eliminate the unknown before the incident occurs by becoming familiar with target hazards and construction features in the areas you serve and protect. If your agency has a rule about the apparatus never leaving the engine house, there needs to be a change. Company training should include routine, informal inspections to all building types, hazards, neighborhoods, subdivisions, and special areas of interest such as rivers, lakes, state parks, and other nature areas. Familiarization is not solely the responsibility of the company officer; it is essential knowledge for *everyone*, including the probationary member fresh from rookie school.

Every time the apparatus leaves the bay floor, crews should take the opportunity to learn about their surroundings. Do more than just walk and look. Pay attention to detail and to features that will affect operations at the next incident. For example, when you enter a structure do you pay close attention to the front door? Sure, if it is a commercial building it is an outward swinging door, probably constructed with tubular metal and glass inserts. But did you take it a step further and look at the locking mechanism? Is it a typical commercial mortise lock and is there a panic type lock directly below it? Don't stop there. How is the lock cylinder inserted? Does it have a collar or is it flush, making it difficult to pull for forcible entry procedures? Observe with a purpose, play the "what if" game, and anticipate situations. Ask yourself questions like, "If I was lost inside this open space with no visibility, how would I get out? What features does this structure have that could hamper operations? Can I easily get a line stretched into the structure in a timely manner?"

Residential preplanning

Preplanning commercial buildings, both formally and informally, is generally not a problem due to common access to public buildings. What about residential buildings (fig. 3–2)? How do we get permission to familiarize ourselves with private dwellings? We can't just forget about it and take it for granted that fighting fires in residential structures isn't as challenging as operations in commercial buildings. Firefighters get lost, trapped, and killed in residential structures more often than in commercial fires. So how do we preplan residential structures? Typically, we will not knock on the occupant's door and ask permission to view their property. In fact this will probably be met with resistance. However, there are ways to accomplish this task:

- Visit construction sites and review construction practices of all new construction.
- Get information from your fire marshal or inspector to all operational staff members in relation to new construction or renovation projects within your jurisdiction.
- Implement inspection programs for all home sales or renovation projects. (This option is not very popular since it is perceived as an extra fee and redundant inspection.)
- Always take the opportunity to familiarize yourself with the structure when answering other alarms at that location. If you respond for a medical call, take the time after the alarm to look at layouts and features that will affect operations (without being intrusive).

- Know the age of the structure or subdivision. This can give you a general overview of construction materials and features. This is very important in suburbia due to the fact that newer construction uses materials that do not hold up well under fire conditions and will affect safety and operations. The time for a safe interior attack will be considerably shortened. However, be extremely careful with making generalizations. Keep in mind that alterations can always be made to structures.

Fig. 3–2. Most homes in a subdivision are constructed with similar layouts and features. (Photo by Erin McGruder)

Generally, homes in a subdivision are constructed with similar layouts and features. So getting a good glimpse in one home will give you a fair idea about the homes in that area or development.

Advancements in construction technology allow builders to replace solid pieces of expensive, heavy support timbers with fabricated wooden joists and trusses. These lightweight, geometric building construction elements may be less expensive, but burn and fail much faster. The failure rates of modern construction materials such as lightweight truss construction, OSB structural materials, and laminate structural members can critically decrease an occupant's escape time or cause critical failure upon a firefighter's arrival.

Currently, New Jersey, New York, Vermont, Florida, and Massachusetts and some cities in Maine, California, Illinois, Virginia, Indiana, and Connecticut use placard systems to identify building construction practices that are hazardous for firefighters in residential occupancies (fig. 3–3). Progressive fire marshals and fire service organizations develop the placards to identify those structural features that are unsafe during fire conditions. The National Institute for Occupational Safety and Health (NIOSH) recommends use of the following alphabetic designations to identify structural components used in a building:

- W— sawn joist/rafter construction, wood members
- I—engineered I-joist construction wood members
- S—steel construction
- T—truss type construction
- C—concrete construction

This is definitely a step in the right direction for both firefighter safety and tactical decision making. Unfortunately, however, it entails a culture change and legislative initiatives for implementation and compliance.

Fig. 3–3. Implementation of residential building placards can raise firefighter awareness of locations utilizing various construction materials.

Elements of Thorough Size-Ups

What are the critical elements all firefighters and fire officers must have in order to make fireground decisions and implement tactics and strategies in a safe manner? The size-up attempts to gather these elements for the most effective, efficient, and safe mitigation of any incident. First and foremost, the company officer must attempt to answer these questions:

- What have I got?
- Where is it going?
- How do I stop it? (What resources do I need and which tactics/strategies do I directly implement?)

Stages of size-up

To answer these questions, collect information during the four stages of the incident:

- Pre-incident
- Receipt of the alarm
- On-scene arrival
- 360° size-up

Pre-incident. The pre-incident size-up is also known as preplanning. Preplanning is an integral part of the information gathering process.

At the time of receipt of the alarm. Once the alarm has been transmitted by the alarm center, all responding members should already be gathering information about the incident. For example, at what time the alarm was transmitted and the time of day could affect the conditions pertaining to the emergency. Also, firefighters should already have a good knowledge of their response areas and be able to visualize addresses with building type, occupancy, and associated conditions (such as the population density, average income levels, and the area's crime rate). Other key points that should be considered at this time are construction features, water supply, and the presence of hazardous materials.

On-scene arrival. A lot of information can be gathered upon arrival at the scene. Gathering information on arrival becomes easier with experience and time on the job. Once your rig hits the street, you should be observing everything. Observe the weather conditions, street conditions, and anything abnormal along the way. Listen to radio traffic and key in on important information that is relayed by dispatch or other arriving units. Once in front of the structure, you should be detail oriented and have a system of complete observation. Observe the building, its occupants, the fire, and its conditions. The only one way to achieve this is with the *360° size-up!* Also remember that scene size-up is continuous and does not end until everyone has left the scene.

The 360° size-up. Size-up cannot be flat and one dimensional. If you pull up without looking at the entire picture, you are going in blind. Effort has to be made by the company officer to see the entire structure—all four sides and the top. A good company officer should attempt to get a glimpse of at least three sides (side A, either side B or C, and the roof line) before even exiting the apparatus. Once exited, the officer should then attempt to get a view of the remaining sides and make more thorough observations close up. But what if it is a large structure, such as a commercial or large residential building? For these structures, it may take more than one unit to complete the 360° size-up, but it still must be completed. Operating procedures may dictate that the second or third arriving should make an effort to initially report to the rear of the structure to complete the full-sided view of the structure.

Tactical hint/trick of the trade

The thermal imager, often considered an interior tool only for helping firefighters conduct operations in zero-visibility environments, can be used for so much more. Technological advances in thermal imaging have made the tool much more manageable and easier to use (fig. 3–4). The next time you conduct a 360° size-up, grab the thermal imager and take a look. This tool will let you see more than is possible with the naked eye. It will allow you to see heat imagery, conduction, convection, radiation, and possibly give you a very good idea where the fire is located and where it is going.

The Radio Report

Upon arrival it is essential that the first arriving unit make an accurate radio report describing the incident for incoming units. It should paint a picture of what is happening and what is initially being done. It should also tell incoming units what actions they should perform next, in relation to

Fig. 3–4. Use of a thermal image camera during size-up can help show fire and victim locations not visible to the naked eye. (Photo by 5280 Fire)

standard operating guidelines (SOGs). Keep in mind, the radio report is not synonymous with the size-up. The radio report is the brief transmission of initial findings from the size-up and actions to be taken.

Remember the elements of an accurate radio report with the memory aid IDEAL:

I—Identify the arriving unit(s).

D—Describe the situation and what you see.

E—Explain your intended initial actions and mode of attack.

A—Assume command of the situation/incident. This does not mean, however, that initial arriving units do not initiate interior attack options. It simply means that command for the operation and tactical decision making has been assumed.

L—Let incoming apparatus know where to go or what their intended actions should be.

Following is an example of an IDEAL radio report.

"Engine 6 is on the scene of a one-story residential with fire showing from one window on side A. Engine 6 will be initiating an interior attack using a 1¾ in. preconnected hand line and assuming command. Have the next in engine lay a large diameter supply line from the hydrant."

Another mnemonic that assists with reporting is BOSE HAS, which is more detailed than IDEAL and is better suited for complex situations. It concentrates on seven points:

B—Building

O—Occupancy type

S—Smoke

E —Exposures

H—Hazards

A—Approach

S—Special

Building. We need to ask ourselves on arrival what kind of building is on fire and transmit that information to incoming companies. What defines a building in relation to implementing fireground tactics and strategies?

- Type of structure: Type I–V
- Size of the building, including length for stretches and area for water flow determination
- Number of stories
- Type of roof

Occupancy type. Occupancy type should always be considered upon arrival. We should never forget our number one priority: *to save lives*. Company officers and firefighters should evaluate strategies with regard to rescue and life safety. Analyzing the occupancy type will put this into perspective and possibly send warning flags to incoming units. The types are

- Residential: single-family or multifamily dwellings
- Commercial
- Mercantile
- Strip malls
- Assemblies
- Churches
- Factories

There are numerous signs that incoming officers should look for to attempt to determine occupancy in structures and dwellings:

- Cars in the driveway
- Handicap equipment and ramps, indicating disabled occupants who may be unable to exit the structure
- Information from occupants who have already exited the structure
- Information from bystanders

Smoke. Reading smoke and describing it to incoming units is the most detailed way to describe a fire in a structure. Remember that reading smoke is predicting fire events, the stage the fire currently is in, and the extent of the fire in the structure. Describing just the fire is not giving the whole picture or anticipating future actions or events. Reading smoke is extremely important for fire personnel and is summarized in detail at the end of this chapter.

Exposures. It is important to indicate which structures surround the structure of fire origin, but most importantly to also detail which buildings are in immediate threat of fire spread.

Hazards. If any special hazards are present, it is essential to note and report these findings. For example, these hazards might include

- Basements
- Power lines
- Hostile threats

Approach. Approach describes having access to the structure. Do you have access to the entire structure? Is the structure set back from the main access point? Another important aspect is having access for ladder companies and elevated master streams. You must communicate conditions that cause challenges to access.

Special. Certain incidents have special situations that are critical to the operations of emergency incidents. Examples of these special incidents include

- Mass casualty incidents
- High life hazards (frequency)
- Water supply (lack of)
- Terrain

Following is an example of a detailed BOSE HAS radio report.

> *"Engine 6 is on the scene of a three-story, Type III ordinary apartment complex with heavy black smoke showing from two windows and fire venting from one window on the second floor. The fire is auto exposing to the second floor with no other exposures directly adjacent to the fire building. Overhead power lines will cause access problems to the Bravo side of the structure, but the aerial will have complete access to the front. There are reports of numerous civilians trapped. Engine 6 is requesting a second alarm and MCI level I. Engine 6 is initiating interior operations and requesting command be set up on side Alpha."*

This is a very detailed radio report and very fitting for the situation. What if the officer of this apparatus was quick to get off the apparatus and initiate operations with a less detailed report? What if he or she elected to transmit an IDEAL report? In this instance, it would be perfectly acceptable for the second due or the first incoming chief officer to relay this information after arrival.

BOSE HAS does not have to be transmitted on every alarm. Not every element will be used; however, remembering the phrase helps to provide a system to cover every possible circumstance that may be encountered at the arrival at an emergency incident.

Post-incident Size-Up

As chapter 13 discusses at greater depth, collecting information from past events can assist with planning for the next anticipated incident. After-action reports and post-incident analyses should detail all findings to improve operations. Firefighters should share information and learn from each incident to which they respond.

Size-Up Mnemonics

A thorough size-up requires a system to ensure all valuable information is collected. In the fire service, many times organized approaches are accomplished with the help of memory aids like mnemonic phrases. There are three commonly accepted mnemonics for size-up procedures:

- WALLACE WAS HOT
- COAL WAS WEALTH
- COAL TWAS WEALTHS

Each covers the basic concept and elements, with a few exceptions. The last of the three is obviously more detailed and includes two more elements. Let's look at the elements in detail and apply them to the suburban size-up and how each one individually affects tactics/strategy.

- WALLACE WAS HOT

    W—Water
    A—Area
    L—Life hazard
    L—Location
    A—Apparatus
    C—Construction
    E—Exposures

    W—Weather
    A—Auxiliary appliances
    S—Special matters

    H—Height
    O—Occupancy
    T—Time

- COAL WAS WEALTH

    C—Construction
    O—Occupancy
    A—Apparatus and staffing
    L—Life hazard

    W—Water
    A—Auxiliary appliances
    S—Street conditions

    W—Weather
    E—Exposures
    A—Area
    L—Location and extent of fire
    T—Time
    H—Height

- COAL TWAS WEALTHS

  C—Construction
  O—Occupancy
  A—Apparatus and staffing
  L—Life hazard

  W—Water
  A—Auxiliary appliances
  S—Street conditions
  T—Terrain

  W—Weather
  E—Exposures
  A—Area
  L—Location and extent of fire
  T—Time
  H—Height
  S—Special considerations

The major differences between the three phrases are the consideration for *street conditions* and *terrain*. COAL considers street conditions, which would easily be detailed as special matters under WALLACE; however, COAL TWAS WEALTH also includes terrain. Terrain is important to the suburban operation and can have a significant impact on tactics/strategies (fig. 3–5).

Fig. 3–5. A thorough size-up system ensures all valuable information is collected. (Photo by 5280 Fire)

Size-up elements applied (in alphabetic order)

Apparatus (trained personnel). The apparatus quantity and type that are detailed for a specific alarm must be considered when choosing tactics and strategies. This consideration is critical for strategic implementation. Remember, this variable can change on a call-by-call basis. It is not a perfect world and your next due engine company might be committed on another alarm. If resources are already scarce, this can be a major handicap.

Apparatus is only half of the variable. The people who are trained to operate the apparatus and equipment can change frequently, possibly hourly, even if you operate a full time, paid fire department. The availability of personnel should not be taken for granted or assumed. If staffing levels change or are inconstant this should be communicated and realized prior to and at the receipt of the alarm.

Area. The area of the structure (length × width) involved has a large impact on operations and decision making. Responding units should immediately consider requesting additional resources for confirmed fires in larger buildings. These buildings should not be a surprise and should have been reviewed through district preplanning.

Auxiliary appliances. Many commercial structures are equipped with standpipes and sprinkler systems. When this is the case, SOGs should reflect tactics and strategies with these appliances included. For example, early arriving engines should be detailed to supply these systems. Keep in mind that many variances and exceptions are made to building codes with the addition of sprinklers. It is assumed that the sprinkler will protect the combustible building features. Therefore, without the sprinkler the building could be defenseless. In this instance, if your operation does not call for supplementing the sprinkler water supply, you are exposed to major negligence and liability.

Standpipes also should be considered in tactical decision making. If a building is equipped with a standpipe system, strategies should involve the tactical use of the standpipe for attack line main water supply. If a commercial building has a standpipe system, there is a good reason for it. The stretch will be entirely too long for the apparatus. Chapter 6 provides more information about the use of standpipes.

Once again, fire officers and firefighters should be familiar with the buildings they protect. The time to identify auxiliary appliances is not at the time of arrival. This should have occurred during walk-throughs, district familiarization, inspections, and preplanning.

Construction. Identifying and being familiar with construction types is essential in structural firefighting. It is obvious that fires behave differently in different types of structures. The construction type can give hints to how fire will spread, the structure's ability to withstand the effects of fire, and the likelihood of collapse. All of these elements impact the decision-making process.

For example, for a working structure fire in a Type V wood frame compared to the same fire in a Type I non-combustible, you should probably be concerned with the rapid fire spread among the wood frame members. The lesser durability of wood members increases the possibility of an early collapse.

When evaluating construction, we should also consider building practices and features, not just type. This is very important, specifically in suburbia where new construction occurs every day. These new building techniques include engineered materials (which do not hold up as well to heat or fire compared to solid elements) and lightweight truss construction. It is important to note these components because it will give you much less time to offensively attack a well-involved interior structure fire. This is a major consideration for occupant and firefighter safety.

Fire officers should also be cognizant of construction features that can increase fire spread and lead to early collapse. Hidden, unimpeded void spaces seem to top the list. Examples of construction features that typically give firefighters problems are balloon construction and story-and-a-half residential structures. Balloon construction allows for open void spaces from the basement all the way to the attic, while story-and-a-half structures allow for hidden void spaces in the attic due to the false knee walls used to square up the room. Another feature is large, un-compartmented areas. These areas aid in building collapse due to rapid fire spread. A proactive fire officer will be able to identify these features within his or her response area.

Finally, in relation to construction, we should be concerned with the roof construction. Is the roof flat, peaked, or bow-string? These answers could affect interior operations as well as exterior. The roof construction may not allow for safe vertical ventilation. A bow-string type roof construction with heavy, progressed fire conditions may lead to a decision to evacuate the building early or prohibit any interior attack.

Exposures. Exposures to the fire building will considerably change the tactical objectives to locate, contain, confine, and extinguish. In many cases, fire impingement on adjacent exposures delays a direct fire attack. To contain the fire means first protecting exposures, which may dictate the initial attack line (fig. 3–6). Exposures may also dictate a greater resource need. If you have exposures involved, you will probably have to call for additional apparatus or multiple alarms.

Fig. 3–6. Exposure control can require a change in tactics and additional alarms. (Photo by Steven Heidbreder)

Height. Building height adds a different dimension to the challenges of firefighting. Fighting fire above grade adds extra risk and requires special tactics. Firefighters, when operating in structures, should have accessible egress points, even when above grade. Also, keep in mind that anything above grade will require more effort when stretching lines and rescuing trapped occupants. Height will affect the placement of aerial devices and their ability to reach upper floors as well. Considerations should be made immediately upon arrival for the placement of ladders before a logjam of apparatus makes the structure inaccessible for essential equipment.

Finally, high-rise operations (regardless of definition) will require four to seven times more resources than operations in single story buildings.

Life hazard. Determining if there is a life hazard should be the most important thing we do, and likewise is the most important element of a size-up. The presence of life hazards will completely alter fireground operations. Remember, the number one priority in the fire service is saving life. Regardless of the structure, we have to assume that life safety is at stake until

1. It has been determined by fire department personnel that the structure is clear of inhabitants.

2. Life can no longer be sustained in the structure or fire conditions make survivability impossible.

There are hints that company officers can watch for to determine if occupants are still in the structure:

- Type of occupancy
- Eyewitness accounts
- Occupant accounts
- Time of day
- Vehicles in the driveway

Location and extent of fire. Before any actions or resources can be committed on the fireground, the location and extent of fire should be determined. A complete size-up attempts to estimate where the fire is located, the size and extent, the stage of the fire, and where it is spreading. This must be done before the first line is deployed! Remember the objectives of locate, contain/confine, and extinguish. How can we accurately contain and efficiently extinguish a fire if we do not know the location? How do we know what size line to deploy if we do not know the extent and progression?

Fire officers must be trained and experienced with reading a fire building. They must be able to determine not only where the fire is, but where it is going. How? By keeping in mind that fire is now and smoke is the future. Upon arrival, officers should evaluate the working structure fire and be able to predict where and how far it may spread. This is done through evaluating smoke conditions. At the end of this chapter we explore how to do this and what attributes to look for.

Size-up is ongoing, therefore locating the fire and extent of the fire condition shouldn't just occur upon arrival. Later arriving companies and command officers should be monitoring both the exterior of the building as well as interior conditions to see whether initial attack tactics are working. If a hard interior attack has been ongoing for 20 minutes but hasn't made any change to the heat or smoke conditions, chances are your efforts are ineffective.

Occupancy. The structure's type of occupancy can be realized upon the time of receipt of alarm. A high occupancy structure should instantly throw up red flags and cautions. First arriving companies should be anticipating rescues and planning tactics for removing trapped occupants. Firefighters should be familiar with the different types of occupancies within their response areas. Residential structures are typically more predicable: most are occupied at night and weekends and empty during the work week. Mercantiles are frequently the opposite: occupied during the work day and empty at night and on weekends. Some factories operate a third shift and are never unoccupied. Social clubs, entertainment venues, and gathering places may all vary operating times. Once again, preplanning is essential for discovering this information.

Special matters. All personnel on the fireground should be aware of special matters or special hazards that could affect operations and firefighter safety. This information can be derived from pre-incident planning, at the receipt of the alarm, during the incident, and after the incident.

Special hazards include (but are not limited to)

- Utility hazards (such as power lines down)
- Egress issues (bars on windows)
- The presence of hazardous materials

Street conditions. Often, street conditions and building access can be taken for granted. Street conditions in the suburbs can greatly differ from those in the urban cities. For instance, compare cul-de-sac and grid systems. In a cul-de-sac, access to the structure might be available from only one direction. If this is the case, it is very important to spot and position apparatus before a bottleneck ensues. Considerations might be needed to align the aerials along the front of the building. Also, there may not be an option for a reverse lay, requiring incoming engines to forward in a supply line.

Keep in mind that many streets in suburban and rural areas are narrower than urban streets or may have overhang challenges. These streets should be preplanned and alternate options should be available, such as smaller and more versatile vehicle responses.

Street conditions can change daily due to weather (snow, flooding, etc.) and maintenance. Conditions should be monitored and well communicated.

Terrain. Terrain is a major consideration and a variable for many suburban and rural fire agencies. Terrain can add many challenges to gaining access to structures within particular response areas. For example, natural barriers such as rivers, elevations, and wooded regions can add to response time and severely limit access to structures. Fire officers must take these circumstances into consideration and alter strategies and tactics. This includes considerations for vehicle size for access, adequate water supply, and lack of structural integrity caused by advanced fire conditions, possibly due to longer response times.

Another concern for suburban based operations is adjoining wildlands, such as nature preserves and creeks. Structures that are built close to woods and wildland areas are at risk to natural cover fires. When responding to natural cover fires in these areas, company officers must consider and anticipate the threat to structural exposures.

Time. Time refers to time of day, week, and year, all of which can present clues and hints to factors that can affect fireground strategy. As stated earlier, time can help predict occupancy and operational patterns for a structure. Time can also affect traffic patterns and response times.

Water. Water supply is a critical consideration and should be immediately considered by responding crews. Ninety percent of the time this involves spotting a hydrant on the map and making the decision to either forward lay a supply line, defer water supply to the second due engine, or to reverse lay the supply line from the fire building to a hydrant. However, there are instances where there is much more to the decision-making process.

Large buildings may require a fire flow calculation larger than the quantity provided by one water main. In these instances, water main grids and preplans will be needed to identify the water supply available for the area. In extreme circumstances, the water provider may have to be contacted to boost the water supply and pressure available.

Another example of water supply challenges is the lack of public water supply mains. In many suburban and rural areas, the public water supply has either not been completed or does not provide hydrants for firefighting. Fire agencies should have these response zones targeted and plans should be established to provide for water supply. This includes spotting the closest hydrant to the region, establishing a water shuttle operation, providing for a larger supply of initial available tank water, and assigning tankers/tenders to the response. Chapter 6 addresses the strategic and tactical challenges of this scenario.

Weather. Weather conditions can have a major impact on both fire conditions and fireground circumstances. Firefighters should be aware of the current conditions and weather forecast before the response is received and be able to anticipate these challenges (fig. 3–7).

Inclement Weather Considerations

Extreme Heat	Requires additional manpower due to exertion
Extreme Cold	Freezes water mains and creates frozen slip hazards
Snow	Access problems and roof weight issues
Wind Conditions	Rapidly spread fire
Low Humidity	Dried out fuels aid in rapid fire spread
Storms	Lightning hazards, lightning strikes and flooding

Fig. 3–7. Inclement weather can impact operational game plans. (Illustration by Fire Medic Art)

Reading Smoke

Reading smoke is a skill that every fire officer should attempt to perfect. When conducting a size-up, it is the tool used to evaluate fire conditions. As stated earlier, fire is now and smoke is the future. Don't get blindsided and only evaluate the visible fire. Every fire is different and smoke does not always react or look the same; however, there are attributes of smoke that can be evaluated to help predict conditions (fig. 3–8):

- Exact fire location
- Fire volume
- Extent and path/speed of the fire spread
- Stage of the fire
- Likelihood of hostile fire events

When evaluating smoke conditions, the following rules apply:

- Observations about smoke are best made from the exterior of the structure.
- Nothing is absolute about fire (there are too many variables).
- *Do not read flames!* Flames can attract your attention and cause you to miss key signs.
- Reading smoke is all about comparing smoke attributes from various openings and cracks.

When sizing up a building, we should view it as a container. Each room or space is a container capable of absorbing a quantity of heat and holding a certain volume of smoke. When a fire originates in a particular container, it will go through growth phases and escalate as long as its needs of heat, air, and fuel are met. The fire will consume the contents of the room as long as sufficient oxygen is present. If the container is closed, there is a very good possibility that the fire will consume all the oxygen in the container and smother itself out. However, many times the container is not sealed and either a door is left open, allowing the fire to spread outside the container or a window will break, allowing the fire to burn freely and consume all contents. The walls, floor, ceiling, structural components, and contents are capable of only so much heat absorption. When they have fully absorbed all the heat the material will allow, the heat will be radiated back. Also, the container is capable of holding only a prescribed volume of smoke, and along with the heat, will begin to pressurize. This condition will create signs and give attributes to the smoke that will help answer the key questions about the fire condition.

Fig. 3–8. The color and velocity of smoke can help firefighters determine a fire's location. (Photo by 5280 Fire)

When arriving on the scene, initial observations should include the following:

- The stage the fire is in. Is it in the early, growth, or late stages?
- The container or fire origin. Determine if the fire is confined to the container or if it has spread. Is more than one container involved?
- Whether the container is still absorbing heat. What is the velocity of the smoke?
- Whether the container is in thermal balance. Is the natural flow of the smoke rising and leaving?

There are four attributes we look for when evaluating smoke:

- Volume
- Velocity
- Density
- Color

Volume. Volume can help determine the size and extent of the fire. The volume of smoke will disclose the amount of fuel that is off-gassing. It will also indicate the extent of the container and whether it is capable of holding any more smoke, hinting to the size of the fire. The volume of smoke from openings outside the container can also help determine whether the container is open (for example, an open hallway door) or if the fire has progressed beyond the container of origin.

Velocity. The velocity of the smoke indicates pressure buildup and can help determine exact fire location. It can also help predict critically hostile fire events. Velocity in smoke is characterized as either *laminar* or *turbulent*:

- Laminar smoke flows naturally and does not have high velocity.
- Turbulent smoke has high velocity and flows violently.

Only volume and heat can cause pressure buildup for smoke to travel with velocity. However, volume-caused velocity will dissipate and slow down immediately upon escaping the container. Heat-caused velocity will slow gradually and remain turbulent. Turbulent smoke can be a good indication of fire location. Upon arrival, fire officers should compare the velocities of the smoke exiting the structure and the openings. If the openings are typically the same size, the smoke with the highest velocity generally points to the fire location.

Velocity is also a major sign for a flashover. Turbulent, high-velocity smoke is an indication that the room (container) cannot absorb any more heat but is radiating heat back. Soon, the container will reach a trigger point where all contents in the room will simultaneously ignite.

Density. The density indicates the amount of ventilation and combustibility of the smoke. The more thick the smoke, the more dangerous it is. Zero visibility smoke indicates that the smoke has fuel continuity, which can ignite without warning.

Color. The color of the smoke indicates the stage of the fire and the distance it traveled from the seat of the fire. The following assumptions typically apply about the color of the smoke:

- White is cooler, black is hotter.
- White smoke indicates farther travel from the seat of the fire. This is due to carbon and hydrocarbons being filtered out during the path of travel.
- White smoke with velocity indicates a hot fire, but at a distance.
- Thin, black smoke indicates fire pushed smoke.
- Brown smoke indicates that unfinished wood products are being heated and the fire has progressed into structural members. This is no longer a contents fire, and could induce a structural collapse.
- Color rarely indicates the type of materials burning.
- White smoke indicates early stages of burning, and black indicates late stages.

There are factors that can influence the four key attributes:

- The size of the container. The size of the container can give different meaning to the attributes. For example, light smoke showing from a large commercial structure like a warehouse illustrates a different volume than that of a residential structure. Do not presume a small volume of fire in the warehouse the same as you would at a residential structure. The ceiling height allows for a greater collection of smoke volume and the pressure buildup will take a considerable amount of time before any velocity can be established. Differences should be noted in structures and not be considered to be equal in relation to size-up.
- Weather. Weather can affect smoke conditions. Different atmospheric pressures can cause smoke to act differently. Heavier,

warmer air can keep smoke from rising. Cooler air can quickly slow velocity.

- Firefighting efforts. Firefighting efforts should make an impact on the fire conditions. If the attack line is reaching the seat of the fire, the fire should be cooling, causing changes to the smoke:
    - Drop in velocity
    - Change in color (dark to white)
    - Decrease in volume and density

  This is extremely important in evaluating the ongoing size-up. If this isn't occurring, tactics aren't working and changes to the operation should be considered.

Reading smoke can help when determining and anticipating hostile fire events such as flashover, smoke explosion, rapid fire spread, and backdraft.

Flashover. The smoke will be turbulent, hot, and very dark. There will be signs of rollover and auto ignition outside container. Remember, once the container is heat saturated the heat is radiated back and all fuels within the box are ready for simultaneous ignition.

Smoke explosion. High-density smoke is an indication for a possible smoke explosion. This smoke is a fuel nearing its trigger point! A brief flash in a pocket of gas reaching its flashpoint will cause a rapid surge of remaining gases. The high density indicates the presence of fuel in the smoke, able to ignite and explode.

Rapid fire spread. High-velocity smoke is an indication that a rapid fire event is about to happen. Hot smoke can flow quickly and rapidly spread fire through hallways and stairwells, possibly trapping occupants and firefighters.

Backdraft. Smoke density can indicate backdraft conditions. A closed container charged with gases at or above their ignition temperature can explode if oxygen is suddenly introduced. This oxygen-deficient container creates a less dense smoke that is almost yellow-brown in color. Other signs of impending backdraft include smoke oozing from seams or cracks; bowed, soot covered windows; and no visible flames and whistling sounds.

Chapter 4

The First Arriving: Primary Decision Making

First Arriving

The initial decisions made by the first arriving officer are the most crucial on any fireground. This holds especially true for a suburban company officer with limited options. Often, preliminary quick decisions such as where to place attack lines and where to vent can make or break the entire incident. Deciding which tactics to undertake is challenging and stressful, especially when you are a single resource attempting to undertake numerous functions alone. Unfortunately, many suburban companies must choose to function as either an engine or truck. In many instances, companies respond from single engine houses and work alone until second arriving units make the scene. Therefore, added pressure is placed on the company officer to make decisions based on these limited circumstances and available resources.

Imagine this scenario: You are the first arriving engine to a three-story, garden style apartment building. Fire is reported in a second floor apartment. You initially see fire showing from one window with heavy black smoke visible from the adjacent windows. It is 3 a.m. and numerous occupants are still in the building. Luckily, you were blessed with a four-person engine for this tour. What are your initial actions?

First, let's analyze the real challenge at hand: life safety. It is very possible that there are numerous lives at stake in a building with a growing/advancing fire condition. But where are those people located in relation to the fire? Do we have time to search for occupants and hope that the fire does not advance, consume more lives, and possibly block our egress above grade? Where and who is our second arriving company?

It is obvious that this company does not have the resources to handle all the challenges initially presented. But where do we start? Where do we commit our initial attack strategies/tactics? We know the highest priority on the fireground is to save lives. So isn't the obvious answer to drop everything and go straight to rescue?

The answer to this challenge is not that simple. Before we go any further with possible solutions, analyze the tasks with which this company officer is faced:

- Perform a complete and accurate size-up
- Establish water supply
- Spot the apparatus in a position to allow for lines to be stretched and aerial apparatus to have access to upper floors
- Take command of the incident
- Decide on attack mode
- Decide on initial attack weapon
- Attempt to rescue occupants
- Stretch attack lines of adequate size
- Place attack lines in the proper position
- Contain/confine/extinguish the fire
- Protect exposures
- Ventilate for fire and lives
- Place ladders for egress
- Assure adequate resources are en route (strike additional alarms if necessary)
- Conserve property

You know this officer cannot accomplish all of these tasks at once. It is simply not possible. Unfortunately, this is exactly where company officers make mistakes. During challenging alarms that tax our abilities more than average, we tend to become overwhelmed, confused, or simply try to do too much. Attempting to do too much with too little can quickly translate into being ineffective to the situation.

Instead of a "jump off the apparatus, realize there is a fire, and quickly rush in" approach, use a logical method of decision making. As a company officer, by combining of all your previous experience and knowledge with situational awareness, current resources, and the capabilities of your company you can derive a decisive initial strategy based on prioritization of impact events. In simple terms: Do the most good for the situation with what you have!

Going back to the example, what is going to make the biggest impact to the situation? In most cases, the answer will be to contain the fire. This is very controversial; however, correct. This chapter explains why. It also explains the circumstances that sidetrack company officers from making initial critical impacts to fireground situations. Discipline is essential!

Underlying Principles

How do company officers make an impact on initial arrival (fig. 4–1)? It goes back to the three basic questions that every company officer should ask:

- What have I got?
- Where is it going?
- How do I stop it?

Fig. 4–1. First arriving company officers performing scene size-up should ask, "What have I got, where is it going, and how do I stop it?" (Photo by 5280 Fire)

What have I got?

The most important thing a company officer can initially perform is the scene size-up. It is such a crucial component in fireground tactics/strategies that chapter 3 is dedicated to the subject. If we do not know all the facts and situational circumstances, how can we perform effectively to impact the emergency? Don't allow yourself to focus on the flame. Realize the situation and see the entire picture. It will save your life and the people you pledge to protect.

Consider some possible scenarios:

- Fire contained in the room/area of origin
- Residential vs. commercial occupancies
- Heavily occupied structures
- Victims trapped
- Heavy fire volumes beyond the capability of hand lines
- Rapid, advancing fires
- Exposure challenges
- Mass casualty incidents
- Inaccessibility
- Unsafe building conditions

No book can possibly detail every scenario. Each day brings new emergencies and situations to the forefront. There is no such thing as a routine situation. These examples are just some of the challenges that are experienced with frequency within this industry. Don't expect the obvious and stay away from the word *routine*.

Where is it going?

Honestly, this question out of the three is missed most often. Yes, the majority of company officers understand the importance of thorough size-ups and conduct them with accuracy. It can also be an easy mistake to march that initial attack line right to the head of the fire and go straight for extinguishment. Eight times out of ten this might be effective, but it is not the answer with a quick, advancing fire. Failing to answer this question is not only gambling the entire structure, but also all surrounding exposures. In certain circumstances it is completely acceptable to write off the building of origin to immediately protect exposures; however, it is important to read the event indicators, such as smoke, to estimate where the volume of fire is headed.

How do I stop it?

Before this question can be asked, the previous two must be thoroughly answered. Do not get flustered and go straight for action without knowing the situation. What is the most common mistake made by company officers? When, regardless of conditions, we defer to deploying the preconnected 1¾ in. hand line, which is almost always set at 125–150 gpm and generically loaded at 200 feet. Ironically, this preconnected generic line was created for the suburban fire department. It allows for quick deployment and in theory can be operated with less manpower. Unfortunately, it becomes relied upon for almost every fire situation, even those past its capability. *Do not fall in this trap*!

As a company officer, you must realize that there are many actions you can initiate to control a situation. You have many weapons at your disposal. You must also realize your inability to completely control or stop every event. Fire might be so advanced that you may not be able to contain it with the resources available to you on your apparatus. You can, however, make an impact by initiating effective tactics and beginning the groundwork for later arriving companies to fill in and supplement efforts.

There are rare situations where you should not initiate fire attack. These might be situations where the first-in engine company might elect to rescue occupants and defer attack operations. Later in this chapter, these rescue situations are illustrated with a matrix and examples.

In the rarest of situations, you might conclude that the option of doing nothing is the tactic of choice. As a company officer, I have been fortunate never to have been faced with a circumstance where the price and risk of actions far outweigh the benefits of the outcome or where mitigation actions only further complicate the event. An example of this would be the unstoppable release of a flammable gas from a container. In this situation, it might be more beneficial for the hazard to be burned off than to let it leak into the atmosphere.

Apparatus Placement

Apparatus placement upon arrival is important for smooth operations. This is especially true of the initial arriving apparatus. Many factors dictate where apparatus should be spotted at a structure fire:

- Access to the building
- Access to the location of the fire

Offensive attack mode

It is essential to attempt an offensive attack anytime life is in peril or if occupants may be in the structure. However, before committing to this tactic, determine whether life can still be sustained within the structure. For example, you arrive to find a one-story, wood frame residential structure with heavy fire involvement (at least 80–90% involvement) and the neighbor advises that she doesn't know if the occupant was able to make egress. Do you still attempt an offensive attack? No, it is futile because the possibility of someone still being alive in the structure and viable is next to zero. Once again, tactical decision making requires discipline. It is a very difficult decision to contemplate that a child who may be in the fully involved room is no longer alive and that attempting a rescue is not an option.

Defensive attack mode

When do we decide to go defensive? There are many variables that dictate a defensive attack mode:

- The structure is so well-involved that there is absolutely no way of saving the building.
- The structure is not stable enough to support interior operations.
- The structure is abandoned (confirmed with no occupancy) and previously been determined unsafe.
- It has been confirmed that there are no occupants in the building, or that no viable life could continue to exist in the structure.
- Exposure risks (containment is necessary) are too high.

Transitional attack mode

Remember, there are situations where you can switch modes of attack and be transitional in your tactical approach. An example I have learned from experience is the attached garage scenario. Often we arrive to find an attached garage 90–100% involved (fig. 4–3). This is a heavy volume of fire! You can try to take an attack line in the front door and try to get ahead of it, but this is a large gamble. For one, you are assuming that you have enough gpm to stop the fire from entering the structure through the garage opening. Secondly, you are assuming that the fire spread is only located on the first floor and not into the second floor or attic. Often the fire has so much velocity and size that this just isn't a good bet. I prefer to use a transitional approach to slow the fire's velocity. Hit the heavy volume of fire

Fig. 4–3. An attached garage 90–100% involved in heavy fire volume. (Photo by Box Alarm Production Fire Ground Photography)

with a master stream or large hand line (2½ in.) from the exterior, while preparing an immediate interior hand line at the front door for interior containment. Don't forget to check occupancy as well!

There is also the transition from offensive to defensive. The officer could arrive and find a slight possibility of a trapped, *viable* occupant. All efforts should be made for rescue. There is a possibility that once this rescue or search has been completed, it might be necessary to pull all interior crews from the structure and move toward a defensive attack plan.

Choosing an attack mode requires experience and discipline. An experienced fire officer should not get excited or lose focus of objectives during a well-involved structure fire. It requires discipline to decide "This is a loss." There are fires that all the aggressive firefighting tactics in the world can't save. A good officer will realize this and go defensive for exposure protection and containment.

Finally, it is extremely important to communicate the established attack mode to all incoming apparatus. Do not make the game plan a guessing game. Get everyone on the same page. If you know you have a large defensive job, communicate it so that other companies can start planning. Aerials can start looking for access to the structure for establishing elevated master streams and engines can start looking for alternate hydrants to establish a large water supply.

Capability

Fire companies should know their limitations. We often fail when crews are pushed past their capabilities. Sounds like an easy concept to understand, right? How many times have you arrived on a well-involved fire and found a three-person crew split to operate two lines? This tactic might work if the fire is on the same grade with a straight shot to the fire. More times than not, both attack lines fail to meet the seat of the fire or are able to move effectively. Or how about the engine crew that splits and now one firefighter is conducting an above floor search while the other crew member is trying to maneuver a hand line to extinguish the body of fire? More is not always a better concept.

First arriving companies are effective when they are able to impact the situation and set the tone for other arriving companies. There is a reason that first alarm or box responses have more than one apparatus assigned. Suburban firefighting needs to accept the fact that five- or six-person engine companies are rare. In fact, we are lucky to have four. That is why we need to depend on a strong game plan of companies working together as a team to accomplish fireground tasks.

I was once detailed to a second arriving engine on a three-story apartment fire. The apartment of origin was located on the third floor. The first arriving apparatus was a quint and only detailed with three personnel for the tour (driver, officer, and firefighter). I was luckily detailed with four personnel. The first arriving officer immediately realized that the fire was two stories above grade and began to the make the stretch. He immediately requested assistance from my company (per SOG) for both the stretch and fire floor truck functions. My crew was able to couple with the first arriving and effectively stretch to the third floor, force the front door, and locate the fire. We were highly successful and were able to rapidly knock down an advancing apartment fire.

What if things had developed differently? What if the first arriving officer had attempted the stretch by his company alone and asked the second company to stretch an additional line? For starters, it would have taken much longer for either company to make the stretch. If you have any experience with apartment fires, you're aware that time is of the essence as fires can move very quickly through an apartment complex. Secondly, there is no way that a two-person hose team could have stretched that line and carried forcible entry tools all the way up to the third floor. They would have gotten to the fire floor and been unable to gain access to the room of origin.

Pick a priority and stick with it. If you decide that you need to make a rescue upon arrival, make the rescue the sole purpose of your company for that initial action. If you decide that containment is the tactic of choice, use your crew effectively to place the line that will make the largest impact for your chosen tactical strategy. As a company officer, you should maximize the potential of your crew by concentrating efforts and focusing on realistic accomplishments. Separate tasks and assign them accordingly. This starts with the first arriving apparatus. Realize that if you cannot accomplish a task alone, ask for assistance and use the resources from other incoming apparatus.

Water Supply

Today's fire apparatus in suburban America is primarily set up to forward lay for water supply. Why is this? There are three major reasons:

- Aerial devices located on engines (quints). If the first arriving apparatus is a quint and the number of aerial devices on the response plan is limited, this apparatus should not be placed on a hydrant. It needs to be placed in front of the structure for aerial access.
- The preconnected cross lay. Suburban departments have become so reliant on the cross lay that we use this tool on almost every structure fire. It had become almost *routine* to pull the apparatus in

front of the structure and deploy this line. This cannot be done if you are sitting on a hydrant.

- Smaller hose beds/less hose. Reverse lays require more hose and different types of hose bed loads. Often they require a horseshoe load with a large diameter static supply to be deployed when the apparatus takes off for the hydrant. Unfortunately, engines are carrying more rescue and miscellaneous tools to compensate for our lack of true heavy rescue companies. This requires more compartments, which in turn takes away from the size of the hose bed. Similarly, apparatus with aerial devices have much smaller hose beds in comparison to straight engines.

A common tactic used in suburban firefighting is for the first arriving engine to have the option of securing its own water supply or deferring water supply to the second arriving engine. There are a few variables that impact this decision:

- Location of the hydrant/water source
- Time differential between first arriving and second arriving engine
- Manpower
- Size of booster tank
- Volume of fire

It is acceptable for the first arriving engine to pass the hydrant and begin initial attack operations using booster tank water. This is done to assure quicker interior operations and maximization of manpower. When the first arriving catches the hydrant, they lose time and a firefighter must remain to make and operate the connection. If arriving with a three-person engine (a driver, an officer, and firefighter), the officer is now left to complete the 360° and begin assembling the attack stretch alone, while the firefighter at the plug turns on the hydrant and rushes back to help with the interior stretch.

The first arriving engine should stop and secure its own water source in some scenarios. The first instance is the case where the second arriving engine will be delayed. If the first arriving has a 500-gallon booster tank and needs a 150 gpm continuous flow, there will be just over 3 minutes of operation time (fig. 4–4). Not a very long or safe duration! A large volume of water—way more than 500 gallons—will be needed if heavy fire volume is noticeable on approach to the structure. For this scenario, you can predict the need for a master stream. The benefits of securing your own water source far outweighs the cost of a slight delay in stopping at the hydrant.

If the first arriving has a 500 gallon booster tank and needs a 150 gpm continuous flow, there will only be just over 3 minutes of operation time. Not a very long or safe duration!

Fig. 4–4. A first arriving engine may need to secure its own water source before initiating other tactics. (Illustration by Fire Medic Art)

Never say never, and do not get stuck on generalities. Some departments still regularly utilize the reverse lay. There are many benefits to the reverse lay if your apparatus is set up to support this. Reverse lays require a sufficient amount of hose to be taken off in front of the structure. Usually, the load is begun with a bundle that can be easily grabbed by one firefighter and placed in front of the door. Typically this bundle is in the form of a horseshoe and is packaged in either 50 ft or 100 ft length to allow at least 50 ft of hose to be arranged at the front door for an easy initial stretch into the interior. Once the bundle is pulled from the truck, the other firefighter(s) should pull off the rest of the amount needed for the stretch. (This is usually estimated using length × width + 50 ft per elevation.) Once the apparatus advances forward toward the hydrant, additional incoming apparatus have full access to the structure. Crews should also remember to assemble forcible entry tools in the front yard before the apparatus takes off for the hydrant.

Truck vs. Engine

Suburban agencies without dedicated truck companies are often asked, "When you arrive with your engine that has an aerial device, how do you decide if you are a truck or an engine?" The answer to this question is simple: *It doesn't matter*. Do not get hung up on the question. Do not focus on differentiating tasks between truck and engine work. The first arriving to a structure fire with a pump faces the same objectives. The tasks that must be accomplished are the same! Suburbia is not a perfect world where you arrive as an engine and your buddy, who responded from the same engine house, arrives simultaneously with a ladder company. Focus on the task at hand.

For the majority of circumstances, first arriving will be placing that pump into service and hopefully deploying the first attack line. This may not always occur because circumstances might present an imminent rescue situation. Rescue functions are typically completed by truck companies in larger, urban settings and so this action is considered a "truck" company operation by most academic standards. Do not let this confuse you! It is not the apparatus type that dictates the action; it is the circumstance.

The Rescue Matrix

When I develop suggested operating guidelines or teach firefighters about fireground priorities, I typically stay away from the acronym REVAS:

- Rescue
- Evacuate
- Ventilate
- Attack
- Salvage

I do this purposely, although the principle is correct, because it is easily confused with a first arriving action list. REVAS is a priority list; it isn't a checklist that can always be followed in order by the first arriving company. Firefighters often reference REVAS when arriving on the scene of a structure fire and immediately conclude that they must rescue (search), regardless of the situation and circumstances. This is not always the proper tactic; the right *priority*, but not the right *tactic*. The reason for this in suburban America is the lack of resources.

The best explanation for this philosophy is that the rescue function may not be as it appears. Traditionally, when we think of the act of rescuing, we either picture a search and rescue team advancing into a structure with a methodical search plan or a crew throwing a ladder to a building to either assist trapped victims to safety or enter the structure to search. In actuality, the most beneficial action a first arriving crew can perform is placing an attack line between the trapped occupant(s) and the advancing fire.

Every situation experienced in the fire service is unique. Nothing is ever uniform or the same. Therefore, we never say the word routine and understand that suggested operating guidelines are just that, *suggested*. The same holds true for rescue situations. The toughest decisions a young firefighter or officer will make are determining the proper to time to defer line placement in lieu of conducting a straight rescue operation with limited resources. The term *limited resources* in this scenario refers to the lack of dedicated truck companies and working alone until additional companies can arrive to assist. Figure 4–5 illustrates this challenge.

Rescue or Line Deployment Matrix

Rescue	Line Deployment
Exact location of victims is known.	Unknown location of victims.
Exact, minimum (1 or 2) number of victims is known.	Unknown number of victims.
Extensive fire conditions.	Large number of trapped victims.
	Unknown location of fire, or it prevents access or egress.

Choose Rescue or Primary Line deployment when not enough firefighters are available to execute each simultaneously and the above conditions exist.

Fig. 4–5. First arriving officers must decide if line deployment is the most effective rescue tactic for the situation. (Illustration by Erin McGruder)

The left column of the matrix describes situations that warrant an immediate rescue without attack line deployment. The right side of the column describes scenarios where the immediate action involves deploying an attack line, even when it is suspected that occupants are trapped in the structure. The main premise for this matrix is the fact that there are not enough firefighting personnel for a simultaneous initiation of both rescue and attack line deployment.

Rescue matrix *(left side)*

There are situations where rescue without attack line deployment is absolutely imperative and mandated. Life safety is our number one priority in the fire service. Buildings can be rebuilt, but lives cannot be replaced. We should never regard property over life and our tactics should always reflect this. When it is within our abilities as an engine company, we must make all efforts to swiftly remove all occupants from harm. The key variables to this circumstance are capability and speed:

- How many victims can we realistically pull from a burning building before we lose complete containment of the fire from its area of origin, egress, or otherwise endanger the entire structure to an advancing fire?
- How quickly do we need to be within the structure? Do we

have time to guess and perform a search without a safety net (attack line)?

It is never advisable to perform functions inside a burning structure without the protection of a hand line. There are situations where we assume the risk to operate without one in the effort to immediately save lives:

- Ladder rescues (assisting occupants from above grade)
- Vent/enter/search procedures

If the exact location of the occupant is known, we can attempt to enter the room from the closest point of egress and perform a very quick primary search to remove the victim. This is a non-traditional search tactic and generally the search will involve only the room that was entered. This tactic assumes a higher level of risk and should only be accomplished by trained and experienced firefighters.

Remember, if you elect to use one of these tactics you are delaying control of the fire; therefore, it is somewhat a gamble. With a rapidly advancing fire, there might only be one opportunity to enter a room before the entire floor of origin is engulfed. Throwing a ladder and entering a structure, although quick, still requires time. Knowing the exact location is critical! If you guess wrong, the opportunity might be lost. Minimize the unavoidable gamble by making informed fireground decisions.

If electing to defer deployment of the attack line, not only should we know the exact location and number of trapped victims, but we should also verify that the resources needed for the rescue are available. Realistically, extracting an unconscious adult takes time and a good amount of effort. Before committing to a course of action, know your crew's capabilities and limitations. If the situation involves multiple victims and numerous locations within the structure, forgoing line placement might save one at the expense of others. Not a very good tactical choice when our goal is to do the most good for everyone involved.

Here are a few classic examples:

You arrive first due to a two-story residential structure. Fire is visible on the second floor from one window. Adjacent to the fire room are two occupants hanging out of a window, and heavy black smoke is also pouring out of the window. They are still both conscious. Bystanders state that no one else lives in the house. The answer for this scenario: order an immediate ladder to the second floor for rapid victim removal. Do not initially attempt to deploy an attack line. The occupants will only have minutes to seconds before the room is engulfed.

Now imagine the situation with a few differences:

> *You arrive first due to a two-story residential structure with two adults frantically screaming and pointing at an overhead window. Fire conditions exist on the first floor. The occupants are the parents of a six-month-old baby who is the only one left in the building. Fire conditions hamper entry into the first floor and access to the stairwell. The answer: conduct an immediate entry into the bedroom window with a ladder raise and conduct a rapid vent/enter/search of the room.*

The last variable for conducting an immediate rescue is extensive fire conditions. If fire conditions are so intense upon arrival that a single interior attack line will not control the fire, complete the rescue of the victim(s) that are located in the only viable locations within the structure. These are conditions where you have only one opportunity to perform any actions before you go defensive. Go straight to the only areas where life may be present and extract occupants immediately. There is no point in deploying an interior attack line in this situation, because the attack line will not make an impact on the fire condition and will have no benefit to any occupant.

I once encountered this situation at 2 a.m. with a single story residential structure. Upon arrival, there was a heavy fire condition consuming the garage and living areas of the structure. Like most single story structures, the bedrooms were on the opposite side of the house from the garage. I was also lucky to have all five of my ambulance personnel on the initial arrival. I immediately ordered a 2½ in. attack line deployed onto the garage (only because my engine lacked a deck gun) and directed two firefighters to conduct a rapid vent/enter/search of the bedrooms. I realized that the attack line was inefficient and unable to make any impact on the heavy interior fire. We were fortunate not to find any occupants in the bedrooms, as the conditions in those rooms quickly deteriorated immediately after our search.

Line deployment matrix *(right side)*

Unfortunately, we may not always have enough personnel to immediately conduct both rescue and line deployment functions. The most beneficial action we can initiate for trapped occupants is to place the attack line between the fire and the occupants; in other words, remove the hazard by extinguishing the fire (fig. 4–6). Initiating the interior attack, for most suburban departments, is the first action in rescue.

Fig. 4–6. No other action saves more lives on the fire scene than the proper size attack line, stretched to the correct location, and placed into service at the proper time. (Illustration by Fire Medic Art)

Does it take time to pull a line, charge it, and place it in the proper position in the structure? Absolutely. However, it is the only answer we have to minimize the hazard or at least place protection between the hazard and the occupants in the building. It also establishes safety for our interior operating crews and hopefully assures a protected egress point for all interior occupants (including firefighters).

In regard to trapped occupants, it is essential to initiate line deployment immediately when one of these circumstances exists:

- The location of victims is unknown
- The number of victims is unknown
- A large number of victims are known
- The location of the fire is unknown
- The location of the fire prevents rescue access or egress

As I stated earlier, there is no reason to take a gamble with an advancing fire. If we do not know the location of occupants, we do not want to search in unknown locations. Do not allow the fire to burn freely, gaining ground on occupants and the structure. The same holds true with numerous trapped occupants. A crew can only physically remove so many victims in a timely manner. The optimal tactic is to begin fire containment and start controlling the hazard, separating it from the victims. This will also set the stage for the incident and allow later arriving companies to filter in and assist with the rescue process.

The location of the fire can also dictate tactics in the rescue. What would happen to a crew who chooses to not deploy an attack line and begins a search for a trapped occupant on the third floor of a three-story residential, only to find a fire is advancing below them and taking over the stairwell? This is a terrible situation to be in and will probably result in firefighters being trapped above the fire, Maydays being called, and forcing deployment of personal bailout systems. It is true that we assume risk when conducting firefighting operations, especially rescues; however, these risks should be calculated and performed with the possibility of saving occupants. Line deployment protects points of egress for firefighter safety and may be necessary to gain access to the interior and trapped victims.

The reality of suburban firefighting, or any firefighting for that matter, is that line deployment should almost always be immediately initiated by the first arriving apparatus with a pump. The reason for this is that we rarely know the exact location of the occupants and exactly how many are trapped. It is extremely dangerous to enter a structure without a hoseline controlling stairwells and protecting points of egress.

Ideally, two crews would arrive simultaneously on a structure and the decision between immediate rescue and line deployment would not have to be made. The engine would simply know to grab the line and the truck would force entry and forge ahead looking for occupants. We know this is not the case in suburban America. So what is the answer? There are variables to consider:

- What is the experience level of the firefighters?
- How many members are detailed to the initial arriving crew?

The most frequent answer is to deploy an attack line, find the fire, place the nozzle where it should be (on the fire), and begin searching back from the nozzle. Remember, we always start from the closest in harm's way and work back. If your nozzle is on the fire, you are starting in the right spot. Additional arriving crews should communicate with the first arriving officer and continue the work and assist with the search. These crews can begin more traditional primary searches once there is a line in place as a security blanket.

For many years, my agency used a picture of an apartment complex fire as a component of the captain's assessment center. The picture showed a large volume of fire in an open, center stair corridor. Numerous people are seen waving from open windows on all floors of the three-story complex. You are assigned as the first arriving engine. What do you do?

The scenario is designed to distract, cause chaos, and challenge the aspiring officer with stress. It is a lot to interpret, but very real. The real solution is to focus on the hazard and to decipher what is really going on. Then ask yourself what are your capabilities and how will you make the largest impact for problem mitigation. The answer is to not allow the victims to distract you and to begin to extinguish the fire with a large attack line or master stream.

If you elect to begin rescue in this scenario, find some answers:

- Where do you begin? Who do you rescue first?
- Do you know the location of all the victims and how many there are?
- Do you have the ability to gain access to victims?
- How many victims can your crew effectively rescue?
- How do you conduct operations safely without egress protection?

If you begin rescue, the scenario will most definitely end in disaster. The fire will advance past the stairwell and begin to consume the adjacent apartments, endangering all occupants. It is also a very good possibility that all firefighters above floors will have their egress blocked and increase the number of rescues when they have to transmit "Mayday."

Weapon Selection

After the first arriving officer completes the size-up and selects an initial tactical strategy, he or she must select the tool or weapon of choice to most accurately achieve the desired outcome. For the fire service, this weapon can be any device able to effectively contain/confine and extinguish a fire. It is a very simple concept; to be effective, fire streams must be able to reach the seat of the fire and deliver a volume of water sufficient to absorb heat faster than it is being generated. Therefore, as a company officer, you must have a complete working knowledge of the capabilities of your assigned apparatus. What weapons do you have at your disposal?

Firefighting apparatus traditionally has the following choices for fire stream delivery:

- Low-volume stream delivery
- Hand line streams
- Master stream devices

Low-volume stream delivery

Low-volume streams typically discharge less than 60 gpm. They include hoses such as rubber booster lines, which are typically ¾ in., 1 in., or 1½ in. diameter. A forestry line is an alternative to the booster hose that has been adopted by many suburban agencies.

The *forestry line* is a single cotton jacketed hose, similar in diameter and flow delivery to the booster. It can be stored in a bundle, a roll, or even flat loaded. The main reason for the adoption of the line is space. Booster reels are large and require a motor. They take up much needed compartment space that may be otherwise used for tool storage. The disadvantage is the forestry line requires more extensive cleaning and cannot be simply rolled back up like the rubber booster hose.

Low-volume hoselines are intended for small exterior fires (fig. 4–7). These fires may include small rubbish fires, dumpsters without exposures, and brush/cover fires. *The lines are not intended for interior firefighting.* They do not provide adequate protection for interior crews and do not meet the minimum of 125 gpm established by NFPA for interior hand lines.

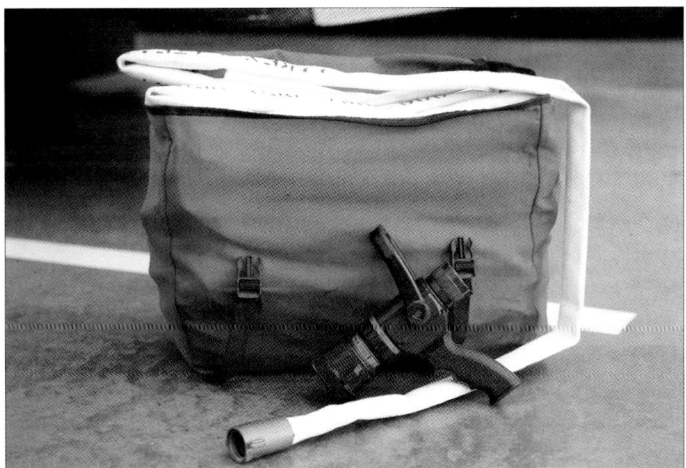

Fig. 4–7. The low-volume forestry line is intended for small exterior fires. (Photo by Erin McGruder)

Hand lines

When discussing hand lines, we must evaluate selection options:

- Size
- Length
- Hose load
- Nozzle (solid or fog)

Typically speaking, a hand line stream is supplied by a 1½ to 3 in. hose that flows from 40 to 350 gpm. The hand line is our offensive weapon of choice and often used for defensive operations as well. Apparatus should be set up to give the fire officer choices in stream delivery. Several factors affect this decision:

- Size and extent of the fire
- Location of the fire
- Versatility of the hand line

The fire officer should first determine the amount of flow required to sufficiently handle the fire volume. Secondly, the length of the hand line should be determined by estimating the setback from the apparatus to the structure, adding elevation (floors), and estimating distance from the entry to the fire.

Finally, the versatility of the hand line should be considered (fig. 4–8). I have found that it is extremely difficult to maneuver a 2½ in. hand line rapidly on the interior of a normal sized residential structure. Does this mean don't consider it? *Absolutely not!* The 2½ in. attack line is an effective knock down weapon for large fires requiring flows over 200 gpm. It will knock it down quickly even though it is not easy to maneuver rapidly, especially in tight spaces and around furniture with minimal manpower. Typically, I favor a quick knock down with the 2½ in. (if necessary) with a 1½ in. attack line deployed by another crew ready to move in from behind for speed and extra versatility.

Examples

1. *You arrive to find a 1,500-sq.-ft residential structure with one room involved on the first floor. It is a normal subdivision lot with minimal setback.* The officer can easily make the decision to deploy the 150 ft, 1¾ in. preconnect and flow 125 gpm. This situation occurs quite frequently. Be careful; this is not always the norm and do not get stuck in the belief that there is a "cure-all."

Fig. 4–8. Various hose loads and appliances should be available on apparatus to accommodate fireground needs. (Photo by Erin McGruder)

2. *The same structure is involved; however, the entire attached garage is fully involved.* Hopefully, for this example your apparatus is set up with a 2½ in, preconnected attack line for quick deployment and knock down. This line is ideal for fires that require larger flows (between 250–350 gpm). It is ideal to pull this line and flow at least 250 gpm to knock down this heavy volume of fire.

3. *You arrive at a strip mall/row of stores with a deep-seated fire in the middle of the one unit. You are unable to determine the exact extent of the fire, but it is a well-involved fire already consuming a good portion of the unit.* The wrong answer for this example is the 1¾ in. preconnect. Commercial structures have a higher fire load than residential and 2½ in. attack lines should always be considered for larger gallon delivery.

4. *You arrive at a 3,000-square-ft residential, two-story structure with fire located on a back, second story bedroom. The house is set at least 200 ft back from the main roadway.* This fire is past the length of the 200–250 ft preconnect. This is a common error; there should be hose loads on the apparatus set up for this stretch. Many engines carry a bundle of 1¾ in. attack line (50–100 ft) attached to a gated valve that is supplied by a static load of 2½ or 3 in. hose on the rear

hose bed. Unfortunately, because many apparatus have reduced sized hose beds due to space constraints and aerial devices, this load is often downsized. To compensate, companies will have a reverse-loaded static supply of either 2½ or 3 in. hose with a gated valve. When they need a long stretch, they will deploy the desired amount of 2½ or 3 in. line and attach a hotel pack or disconnected preconnect (1¾ in.) to the gated valve. Remember that this requires time and manpower!

5. *You arrive to find a five-story hotel with a fire located in a unit on the fourth floor. The building does have a standpipe system.* For most agencies who have preplanned their response area, this should be a no brainer. The officer will most likely elect to deploy a high-rise bundle and attach to the standpipe on the third floor (one floor below). This pack is usually 150 ft long and ideally 2½ in. diameter.

Typical engine company hand line choices:

- 1¾ in. preconnects
    - Cross-lay
    - Bumper loaded
    - Hose bed loaded
- 2½ in. preconnects

    I will make one quick comment about the 2½ in. attack line in regard to nozzle selection. I have found through experience that fog nozzles make flowing a 2½ in. attack line more of a challenge, especially in limited personnel situations. The solid stream nozzle should be the nozzle of choice due to less nozzle reaction. It also allows for greater reach and penetration.

- Static bed loads (1¾ in. bundles attached to gated valve, supplied by 2½ or 3 in. static hose)
- Hotel/high-rise packs (prebundled 1¾ in. or 2½ in. attack line)

    Once again, the nozzle of choice for this setup should be the solid stream nozzle due to lower designed system pressures and possible debris in the standpipe, which could clog a fog nozzle. Remember it only requires 50 psi at the tip to operate the solid stream, compared to the 100 psi for the fog stream nozzle.

Finally, foam is the one final consideration to deliberate upon in relation to attack lines. Not all fires can be extinguished with just water. At times, multiple classes of burning materials may be involved. The classic example of this is the attached garage. Besides the typical class A material involvement, there is a strong possibility that class B materials could also be burning. Gas cans for lawn mowers and the fuel tanks from automobiles parked in the garage can pose a challenge. A foam solution may be required for extinguishment. Luckily, modern fire apparatus have the option for a prepiped foam line and an inline foam induction for rapid foam delivery. I am very fortunate that my engine has one of these lines on a preconnected bumper cross lay (fig. 4–9).

A foam line should always be the main line of choice when dealing with a fire with any possible class B fires, such as auto fires. In fact, when I elect to deploy a 1¾ in. attack line, the foam cross lay is always the line of choice because it also gives the option to flow a 1% foam solution for greater smothering effect, even with class A materials.

Fig. 4–9. Preconnected bumper cross lay on Rescue Engine 3344. (Photo by Erin McGruder)

Basic Concept

The basic concept should be embraced by all first arriving company officers. "No other action saves more lives on the fire scene than the proper size attack line, stretched to the correct location, and placed into service at the proper time."

There are critical elements to consider:

- What size hand line and how much flow? The ability to make this decision is developed through experience. Fire officers must be able to relate scientific calculations of heat absorption and apply them quickly on the fireground. Does this mean that each fire officer should carry a calculator in the cab of the truck? Probably not. A good fire officer will use experience and preplanning to develop a rule of thumb, and be able to distinguish a 1¾ in. fire from a 2½ in. fire.
- Where do I place the line? By conducting a thorough size-up, the fire officer should be able to determine where the fire is and where it is headed. Then he should be able to develop the tactic to confine/contain and extinguish the fire.
- When? The answer should always be *quickly* or *now!* Fire will not wait and it will keep consuming rapidly until it is stopped. The timeliness of the line placement could be determined by versatility, access, and available personnel.

The outcome of the fire can be determined by the first few minutes of the incident and the actions of the first arriving crew. "As goes the first hand line, so goes the fire."

Have a plan in place for line deployment prior to arrival. This entails having preassigned riding positions in order to maximize personnel. As I mentioned earlier, placing a hand line can be challenging. Ideally, the following would occur on every fire:

- The hose team concentrates on placing the hand line and delivering a stream to contain/confine and extinguish the fire. This would require hands free (no tools) placement of the line.
- The engine officer has the ability to direct, observe, and communicate the situation (fig. 4–10). In order to be able to accomplish this task, he or she would not have to serve as a backup to the nozzle firefighter. In my opinion, this is the most dangerous habit/activity that we conduct on the fire scene. When the officer is occupied with pulling hose and is working heads down, full

situational awareness is impossible, as is the ability to direct or communicate effectively.

Fig. 4–10. The engine officer has the ability to direct, observe, and communicate fireground situations. (Photo by Erin McGruder)

- The hose team would consist of a nozzle firefighter, a backup to the nozzle, an officer, and a kink chaser. For some difficult stretches (above/below grade, long stretches, etc.) a kink chaser may be required on every bend or corner to assure a smooth stretch.
- The first arriving officer should communicate with the second arriving and direct which actions need to be initiated by the

second arriving (initial line placement assistance, search and rescue, forcible entry, ventilation, etc.).

- The hose team should have a facilitating crew alongside, with tools in hand, for fire floor truck operations (opening up walls, forcible entry, pulling ceiling, breaking windows, etc.).

Remember, this is an ideal situation. I realize this is highly dependent on personnel and apparatus arrival times. Each agency may have a unique situation, but continue to strive for this ideal operation. Developed suggested operating guidelines will move your operation closer to this goal and achieve consistent fireground practices.

An example riding assignment list for first arriving:

- Officer (responsible for all supervising activities)
- Engineer (responsible for operating apparatus)
- Nozzle firefighter (responsible for pulling hand line, placing 50 ft of hose at the front door, discharging nozzle, expelling air and checking gpm and stream pattern, placing nozzle, and operating nozzle)
- Backup firefighter (responsible for assisting with hand line stretch, following line and removing all kinks, monitoring bends/corner; may make hydrant connection in the event of first arriving securing water supply)

Master streams

Some fires require more than hand lines. They will require massive amounts of water or there may be exposure challenges (fig. 4–11). Once again, a good fire officer will understand the apparatus's attack abilities and realize which devices are available to effectively protect exposures and contain large fires. A *master stream* is considered any stream that can deliver more than 350 gpm.

Master stream devices include

- Portable monitors
- Deck guns (mounted monitors)
- Elevated streams (both straight aerial appliances and ladder tower appliances)

Fig. 4–11. Some fires require massive amounts of water that cannot be effectively delivered through hand lines. (Photo by David J. Dubowski)

When would a fire officer elect to deploy a master stream?

- Defensive operations
- Heavy fire conditions
- Exposure problems
- Unstable structures

There is one major caution about the master stream: *Do not flow master streams while personnel are operating inside the structure!* This is extremely dangerous for a few reasons. Master streams can push fire onto interior crews, add to the instability of the building, and the force of the stream can physically injure interior crewmembers.

I have observed a common mistake with aerial master streams. On many occurrences, the first arriving aerial officer has arrived to find heavy fire in an attic and has elected to deploy his elevated master stream onto the attic instead of aggressively attacking from the interior and pushing up and out. Unfortunately, a key concept for firefighting is that it is ideal to push fire up and away from unburned areas, as opposed to pushing it down. Just because the aerial possesses the ability to deliver a large amount of water quickly, it isn't always the best answer. Many attic fires can be hit hard from the interior, saving a large portion of lower floors. The fire is already vented and pushing toward the sky. When flowing the aerial down onto the structure, we are in fact pushing fire into the structure and adding a large amount of water weight that will further reduce stability. I am not saying the answer is always an aggressive interior attack, especially when it has been determined that the structure is unstable or there is no way of stopping the fire from advancing onto other floors.

Master streams add a large amount of stress to the stability of the structure due to water weight and force. Keep this in mind if you are considering a transitional fire attack from defensive to offensive. Elevated master streams, especially, can cause failure with structural members and weaken structural integrity.

Finally, if you determine the weapon of choice to be a master stream, it is essential to quickly secure a permanent water supply. Luckily, fires that justify master streams can usually be seen before arrival. The officer should spot the hydrant before arrival and always have it in mind as an option. If a large volume of fire is visible, grab the water source unless the next arriving company is on your back bumper.

Further Needs

As the first arriving apparatus, you are in charge of the entire incident until the arrival of a chief officer or the next arriving apparatus (dependent on your operating guidelines). Do not delay getting help if you need it. If you arrive and realize situations present will dictate more resources, call for them. This requires knowing the limitations of the initial alarm response.

For example, as a company officer you should know from experience how many attack lines a first alarm assignment can support, barring special circumstances. These circumstances may include rescue, forcible entry, ventilation, access problems, and other functions. My agency's first alarm assignment details five apparatus. This will support one attack line and one backup line with supporting all other fireground functions effectively, including RIT. I know that I will most likely call for additional resources (apparatus) if any of the following are present:

- Special circumstances (as mentioned above)
- Heavy fire that dictates more than two lines for operation
- Difficult stretches, such as commercial occupancies
- Exposures
- Extreme weather that will create firefighter fatigue
- Multiple casualties

We have written into our suggested operating guidelines that the first arriving apparatus or chief officer has the initial option to upgrade the alarm with either one or two additional apparatus or a full second alarm assignment.

Communicate, communicate, communicate!

Make a size-up, commit to operational decisions, and communicate with incoming apparatus. If the next in apparatus does not know the game plan, its crew may not initiate vital tasks to support the desired tactical actions. Suburban fire companies cannot complete tasks alone. We need facilitating and support actions. The first arriving officer must communicate what needs to be done to support the initial tactics, which often entails facilitating the first attack line.

Chapter 5

The Second Arriving: The Facilitator

Second Arriving

The largest difference between urban and suburban operations is the use of truck companies. These companies are intended for the purpose of supporting all suppression activities, but do not involve the direct action of deploying hand lines and flowing water. Because suburban operations most often do not staff dedicated truck companies, operating guidelines and action plans should be established to compensate for the lack of these support functions. This has been emphasized numerous times and is the underlying theme of this book. The second arriving company is the pivotal apparatus that, for many suburban agencies, fills this support role and is detailed to these facilitating functions.

The second arriving apparatus provides a critical component by facilitating the tactical strategy that was initiated by the first arriving apparatus. It is responsible for coordinating and completing numerous functions on a fireground. As stated earlier, a single company—especially one with minimum staffing—cannot be expected to complete more than a few functions alone and will often require the assistance from supporting companies. While the first arriving company's decision making bears a significant impact on the incident, many times these initial actions are fruitless and unsuccessful without the supporting actions from other companies. The actions of a second arriving company represent the true purpose and concept behind *Suburban Fire Tactics* and the need for a systematic action plan or suggested operating methods (SOGs).

Before I had the idea to write this book, I was tasked to reevaluate my agency's SOGs for structure fires. We were doing many things right in terms of structural firefighting, yet we realized there was room for improvement. For the majority of incidents everything went very well, but occasionally something would slip through the cracks and operations would not go exactly to plan. This is not acceptable in the fire service. We realize that every outcome to every fire cannot always be as desired; however, we still want to strive for consistency. After analyzing numerous fire responses, my colleagues and I were able to pinpoint one of the major problems that detracted from consistent operations. The majority of issues involved the use of our second arriving apparatus.

In our SOGs the second arriving apparatus, which for our agency always has a pump, was always detailed to deploy a secondary or backup line, which is essential on all structure fires. Unfortunately, we were missing many other functions that should have been prioritized ahead of this action. We discovered that the structure fires that were not controlled in the desired timely manner, or not controlled at all, had the same reoccurring theme: the first line was not effectively placed in the proper position in time. What caused this? We found many answers to this dilemma:

- There wasn't enough manpower to effectively make the stretch.
- Crews could not locate the fire in a timely manner.
- There wasn't adequate ventilation to facilitate the movement of the attack line.
- Crews were not able to both stretch the line and to open up the walls and ceilings to fully extinguish the hidden body of fire.
- Crews were distracted by attempting to complete too many functions, such as search and rescue.
- Adequate water supply was not established.

The placement of an initial attack line is the most critical function in the control and mitigation of a fire and makes the greatest impact on a fire scene (fig. 5–1). There are many subsequent and simultaneous actions that must transpire for this to occur. These actions must be accounted for and take precedence over less time sensitive functions that provide a smaller impact. Therefore, it can be established that all initiative and effort should be made toward placing the initial attack line before considering deploying additional lines. While these additional lines are essential, the effective and efficient placement of the first line should be the priority. I do realize that nothing is absolute and that there are situations that may call for a variation to the game plan; however, this is the common rule.

Hey, let's get this one in place first.
I may need some help.

Fig. 5–1. Effort should be made toward placing the initial attack line before considering deploying additional lines. (Illustration by Fire Medic Art)

After careful observation, we realized that these facilitating functions were not getting completed in a timely manner. That more often than not, the first attack line was taking a beating and not meeting the objective while the second attack line was behind them doing absolutely nothing. Many times we would find numerous lines on the ground, yet none could reach the seat of the fire. One example in particular was a basement fire. Three lines were deployed, yet none of the crews were able to make the push down the stairs and meet the seat of the fire due to heat, inadequate ventilation, a difficult stretch (numerous bends), and overcrowding (numerous lines in one place and firefighters stepping all over each other).

Each agency has different needs. There is no doubt to what the first arriving company should accomplish. The responsibilities of the second arriving completely depend on what still needs to be accomplished to place the initial line. If you are certain that your individual companies can place the line and complete all facilitating functions without excess burden, then maybe it is in your best interest to deploy the second line by the second arriving crew. However, for most suburban agencies, this is highly doubtful.

I am absolutely certain that there are structure fires that take very little effort to complete a size-up, stretch a line to the first floor, and quickly knock down a fire; these instances require minimal facilitating functions. But tactics should be kept consistent. If you detail your second arriving as a facilitator, why not keep it consistent? Still allow them to complete all the additional tasks without pulling that second attack line. Chances are, in an example like this, the second attack line will not make a large impact to the incident.

The Secondary Size-Up

Any decent urban department truck company officer will tell you the value of a secondary truck size-up. A truck size-up is in addition to the first arriving engine company's findings and report. The engine company (or for our situation, the first arriving apparatus) completes a size-up to determine the building type, occupancy, circumstances, and fire conditions in order to initiate tactical strategies. The supporting company that is facilitating the operation should conduct the same to determine which actions the crew will initiate in support of the engine or first arriving.

The second arriving apparatus can facilitate or support the initial arriving apparatus through many functions. The officer should select the actions that will make the largest, immediate impact to the operation that are needed to facilitate initial tactical decisions. For the majority of incidents, this initial action will be the deployment of the primary attack line.

Findings that could influence this decision making include the following:

- **Water supply.** Has an adequate, permanent water supply been established for the operation?
- **Entry/egress.** Are there entry/egress challenges? Is forcible entry/egress needed initially? Is ladder placement essential?
- **Occupancy.** Is rescue a possibility? Is there a possibility for multiple occupants?
- **Building type.** Is there a possibility for fire spread through hidden voids? Do we need to open up walls/ceilings immediately? Are there any specific construction features that could complicate fire

spread and concealment, including balloon frame construction or knee walls (creating hidden voids)?

- **Fire location.** Do interior crews need immediate assistance with fire location? Is it hidden? Is it a difficult stretch above or below grade, requiring additional staffing for the stretch? Is it above grade, requiring immediate ladder egress for interior crews?
- **Ventilation.** Has the fire vented? What are interior heat conditions for operating crews? Is immediate ventilation required for life or fire?
- **Safety.** Is the interior crew operating under safe procedures and conditions? Can they get out if conditions worsen?

Assigned Duties

Simply stated, the second arriving company's sole purpose is to support all tactical actions taken by the initial arriving company. Typically, the initial truck (facilitating) company functions are categorized as *fire floor functions*. These are functions that occur directly on the fire floor and directly affect life safety (search and rescue) and extinguishment operations.

What actions are necessary to support the attack line?

- Water supply
- Complete forcible entry
- Assist with the hose stretch
- Assist with locating the fire
- Overhaul during fire attack, such as opening walls and ceilings to find hidden fires
- Ventilation
- Search and rescue
- Providing egress (ladders)

Water supply

The first arriving apparatus has the option of deferring water supply and beginning the fire attack using booster tank water. If this is the case, it is imperative that the second arriving secure the water source and provide a supply line for the attack engine. Know that this function must be completed with expedience! If the tank size of the first arriving engine is 750 gallons and the intended flow of the attack line is 150 gpm, the interior attack line only has 5 minutes of flow time. There will be a lag time for stretching the line to the door, but do not lark about. It is a very risky tactic to flow the initial attack line off tank water; however, it is very common in suburban operations.

Complete forcible entry

It is a very obvious concept that if you cannot get access to the fire structure, nothing can occur. First arriving companies should not wait for the second arriving to arrive to begin forcible entry. Forcible entry should be initiated immediately after the size-up and ideally simultaneously with the line being deployed from the apparatus (fig. 5–2). In cases where the first and second arriving occur in close proximity, the first company should concentrate on the initial size-up and attack line deployment while the second should focus on water supply and gaining access to the structure. Allow the hose jockeys to have free hands for the hose and allow the facilitators ("truckies" or second arriving) to handle the forcible entry tools.

Forcible entry does not end after gaining access to the structure. Crews should also worry about forcible egress during the entire incident while conducting interior operations. Things do not always go as planned in horrible situations; interior crews may have to locate an alternative route for egress. Unfortunately, these paths may be locked, blocked, or nonexistent. Because it is more effective for hose team members to have hands free of tools and so they can concentrate on moving the attack line, they need to have a companion crew available with tools that can open paths if needed. Also consider the fact that the front door is not the only door that should be forced. Interior working crews should have at least two egress points, if not one on every side and every floor of operation. Therefore, facilitating crews should complete forcible entry at least on the rear of structures and anywhere there is an obstacle for egress.

Fig. 5–2. Forcible entry does not end after gaining access to the structure. (Photo by 5280 Fire)

Assist with the hose stretch

This function is very dependent on the staffing of the first arriving apparatus and the location of the fire. If the normal staffing of the first arriving apparatus is four, you can deduct that the crew has pump operator, an officer, a nozzle firefighter, and a backup firefighter.

Under normal circumstances, this crew should be able to stretch an attack line on the same level floor (grade) with efficiency. However, there are circumstances that could require more personnel to stretch the line into position:

- Above grade stretches
- Below grade stretches
- Obstacles
- Extra corners or bends

In these circumstances it is imperative to lend staffing to the initial crew to support the stretch.

Assist with fire location

Fires present themselves in all different manners. Remember to look for clues of the fire's progress, location, and intensity in the attributes of the smoke exiting the structure. Sometimes this isn't as easy as it appears. There are times when we get very lucky and can see the fire from the exterior or we can find it immediately upon entering the structure. However, there are times when the fire is hidden and it isn't that easy. Examples of where fire can be hidden include interior rooms, below grade, attics, ceiling spaces, walls, and any hidden void spaces.

For many fires in the growth phase, there can be a very small window of time to locate the fire, contain/confine, and extinguish it. Often, interior attack line crews will require assistance in speedy fire location. Once again, each attack line crew should have companion crews that have the abilities and essential hand tools to assist in this endeavor. It would be very beneficial for a hoseline to have the assistance of an extra officer or firefighter leading the way with a thermal imager. The thermal imager is an invaluable piece of technology, and when used correctly by a trained staff member can lead crews quickly toward patterns of convection, radiation, and conduction and expedite the location process.

Overhaul

Overhaul should not be viewed as that function that only occurs after the fire is placed under control. Overhaul and the act of searching for fire are synonymous. Overhaul specifically refers to locating fire in hidden spaces. Overhaul often must begin during the fire attack in order to completely confine the fire and halt spread. These hidden spaces can include

- Partitions
- Walls (especially walls without fire stops and those of balloon frame construction)
- Knee walls (often found in story-and-a-half residential structures)
- Floors
- Ceilings (both drop and drywall)
- Attic spaces

Hand tools are essential! They should be assigned and carried by each member who is not operating a hoseline. As an officer, I try to always carry a 6 ft hook when entering a residential structure fire. I use the hook to check all void spaces, above and side to side, to avoid fire getting behind my crew, causing entrapment. I have seen many structure fires escalate past being handled by one attack line because crews did not have tools or the correct tool assignments to effectively stop the spread of fire through walls and ceiling members. How many times have you responded for a lightning strike and one of your later arriving partners did not have the forethought to bring a ceiling hook? Why do you want to give the fire time to burn longer above you while you take the time to go back to the apparatus to retrieve a hook? Simple mistakes like this can win or lose an incident.

Ventilation

Ventilation is conducted for various reasons and is an important component of a coordinated fire attack. Ventilation can effectively

- Improve survivability conditions for trapped occupants by replacing smoke and heat with fresh air. (Be careful, fire likes fresh air too!)
- Attempt to change the velocity and direction of an advancing fire by providing a path of least resistance.
- Improve interior conditions for interior attack crews by removing heat and increasing visibility.

Ventilation is a very controversial issue in the fire service because it requires skill and the ability to anticipate the outcomes of your actions (fig. 5–3). Incorrect ventilation—in the wrong place or the wrong time—is the equivalent of throwing gas on a fire. Remember that one of the major components of fire is oxygen. If you give a starving fire fresh air or feed an advancing, growing fire more fresh air, things are going to escalate quickly. When we vent, we must predict what is going to happen to the fire and where it is going to go. Secondly, we should have attack lines already in place to control further growth and spread.

There are many different forms of ventilation. Typically, ventilation is characterized by the intended directions or orientations: horizontal and vertical. Horizontal ventilation attempts to remove heat and smoke horizontally across the same plane or floor. This form of ventilation can easily be accomplished by finding the general room of origin and breaking the window in proximity, allowing the heat and smoke to exit the structure through natural air currents. It can also be accomplished mechanically through positive pressure techniques using a ventilation fan. If your agency uses this practice, you should be trained and have a complete understanding of how you are affecting the fire and which direction you are pushing it. Vertical ventilation is the practice of cutting ventilation holes in the roof of the structure to release the smoke and heat vertically.

Fig. 5–3. Ventilation is conducted for various reasons and is an important component of a coordinated fire attack. (Photo by 5280 Fire)

Vertical ventilation is effective for upper floor fires and fires involved in attic spaces. It does require more time to gain access the roof and to adequately cut a proper vent hole. Because of this time and resource requirement, it is often conducted by crews arriving third or later to complete additional truck company functions.

Search and rescue

Search and rescue is the most important thing we do as an industry/ service. As discussed in chapter 4, "The First Arriving: Primary Decision Making," it is not always the first action that we initiate. In many cases for suburban companies the first action of search and rescue is the initial attack line deployment. That is why it is essential for the second arriving company to further facilitate the attack line (if necessary) and begin all searches in the proximity of the fire and remove those in greatest harm. If the line is placed, the second arriving can search from the nozzle back and can also attempt to find the areas of high probability, such as a home's bedrooms when the fire happens in the middle of the night.

When there is a report of a possible trapped occupant, this function should take the priority over the other truck functions that the second arriving is responsible for. Do not just consider the search from the nozzle back as the only search option. Consider areas of high probability and conduct quick primary searches. Also consider vent/enter/search. Chapter 4 stated that the first arriving company should use the primary search only under very select circumstances. The vent/enter/search tactic is ideal for the second arriving company if the first has already deployed a line.

Providing egress (ladders)

When operating above grade, consideration should be made for interior crew safety and egress (fig. 5–4). Things can change and escalate quickly in a structure fire. Because of this, backup plans should be in place to rapidly exit in case of danger. Although this action is not typically considered a fire floor action or prioritized high among the truck company responsibility list, it is an action that must occur.

How many firefighters does it take to place and extend a ladder? This is no joke. Often, this duty can be assigned to the engine operators who have already established water supply and pump pressures, or who are not involved in any water operations at all. Maximize your staffing potential, especially when it is limited.

Fig. 5–4. Consideration should be made for interior crew safety and egress. (Photo by Box Alarm Production Fire Ground Photography)

Communication

Besides conducting a secondary size-up, how does the second arriving company know which function to prioritize and initiate to help facilitate operations? The answer is communication. Do not play the guessing game. Upon arrival, the second arriving officer should immediately communicate with either the first arriving officer or chief officer, face to face or through radio communications.

There are important facts to know:

- Has water supply been established?
- Have you made entrance or gained access?
- Do you know the location of the fire?
- Have you made progress with deploying the attack line? Are you at the seat of the fire or will you need assistance?
- Are there any reports of occupants trapped? If so, what are their possible locations?
- What are interior conditions like? Is ventilation needed or has it been initiated?

Another important aspect in relation to SOGs is to communicate completed tasks to either command or the later-arriving company that will also assume facilitating (truck) functions. It is important to indicate which functions still need to be completed, such as the following:

- Ladders
- Vertical ventilation
- Additional forcible entry
- Salvage
- Utilities

Making decisions as the second arriving officer in an operation can be as challenging and critical as the first arriving apparatus. Many actions initiated by the first arriving apparatus will require assistance and further facilitation. Remember, in suburbia we do not have the luxury of an overabundance of staff. We have to make our operation effective with the circumstances that we are given. The second arriving company's decision making reflects this concept and is a prime example of the resourcefulness that is required for suburban operational success.

Chapter 6

Special Considerations: Suburban Water Supply

This chapter was contributed to by Byron Long, assistant chief with the Boles Fire Protection District, Franklin County, Missouri and engineer (retired) with the Metro West Fire Protection District, St. Louis County, Missouri.

Suburban Water Supply

In basic fire training, we are taught the inner workings of water supply systems. Modern water supply systems are made of distributors, circulating feed hydrants, grid systems, and other components. Understanding how these systems interact with our suppression activities is extremely valuable, at least where an established water supply system exists. As we move farther away from the big cities, the likelihood that the water supply includes firefighting service components or even exists in the area to be protected is questionable. Suburban/rural sprawl often outpaces the water supplier's plan to provide adequate services. Simply adding hydrants does not suffice where suburbia interacts with rural water systems that were only designed to supply potable drinking water.

Even when water systems are adequate and code dictated hydrant spacing is applied, the types of sites ranging from tract style housing to custom homes with deep setbacks and narrow lanes requires us to plan for the water supply options that are available. Given a typical residential scenario in a properly supplied area, many fire officers can devise a quick attack that will save both life and property. Place the same structure in an area with limited or no water and some of these same people will surrender to the thought that an attack strategy cannot be applied, or worse that the structure will be a total loss.

To win this game requires us to know our needs. Determining our extinguishment requirements can help us adapt our resources to best meet that challenge. A good first start is to supplement your abilities with automatic mutual aid connections. Resist the urge to wait until you are on the scene before requesting aid that may require more time to arrive than your water supply can withstand. Manage your fleet and personnel to achieve your initial attack needs—what you must have to effectively stand alone until help arrives (fig. 6–1).

Fig. 6–1. This structure is located at the end of a 3,500 ft single lane road without hydrants. Are your initial resources able to implement an adequate, sustained attack? (Photo by Byron Long)

Preplanning for Needs and Resources

Rescues, aerials, and extra command staff are excellent tools, but all are for naught if you have not initially established an adequate, sustainable water supply. Cross train or maintain skill sets for those individuals who might be required to operate in uncommon roles for their position. Consider redirecting command and line personnel to staff other apparatus on certain responses as needed. If a critical water supply apparatus may occasionally go unstaffed, have a plan in place to ensure that someone will get that vehicle out the door in a timely fashion.

Many agencies spend time preplanning target and high hazard occupancies. Should we not consider doing the same for properties in areas that have limited access and/or water? It could be quite difficult for agencies that cover large areas or those with limited resources to develop action plans for each structure. A general SOG covering water and supply requirements that will be needed for an incident in those areas of concern could be developed, though. And it can be tailored to the department's particular circumstances.

Apparatus Design Considerations

When specifying new apparatus, be specific. National Fire Protection Agency (NFPA) minimally compliant is just that—a *minimum*. NFPA specifications may fall short of your complete, real world needs. NFPA only requires 250 gpm for tanks under 500 gallons and 500 gpm for tanks with capacity over 500 gallons. Instead of stating in your purchasing bids the 1901 standard, specify the tank-to-pump flow you wish to achieve, and make it a requirement of prepurchase testing.

When designing a new apparatus, take the time to research new suppression technology and, where applicable, add it to the arsenal. That said, no matter the effectiveness of the technology, base your initial response as if you only have water and prepare to operate if the technology fails.

Apparatus required to pump or to supply water in these situations should be designed with such considerations as a top priority. Initial response apparatus should have enough water onboard to sustain needed fire flow until a secondary source is established. This may require engines to have larger tanks than your current standard. Also, specify the optimum tank-to-pump flow that your operations will need.

If tank-based operations must be sustained for extended periods, you should review your current means of tank refilling. Refill methods vary from discharges directly from the pump to manual or automatic fills that connect directly to the tank or are plumbed to auxiliary large-diameter hose (LDH) intakes.

A gated LDH connection, with a combined automatic tank fill option, can be quite convenient if you expect to transition using your original supply line. This can be handy where intake pressures may fluctuate, such as when using a multi-tanker nurse shuttle. Intake pressure fluctuations can be avoided by shunting all incoming water to your attack apparatus' onboard tank. This does limit the flow that your tank-to-pump line can provide, but most tanker-direct supply situations will be limited gpm. The advantage is that no connections need to change once a higher volume system is in place. The pump operator merely redirects the flow of incoming water.

If limited access is a concern, consider using auxiliary intake or suction positions. Auxiliary intake or suction positions work well when spotting a hydrant at a front or side mount connection, but problems can occur on a narrow lane with the water supply behind the apparatus.

Placing a properly sized intake on the back step may speed deployment of secondary water supplies or long forward lays (fig. 6–2). If drafting from a remote position, front or rear intakes will be required. Those intakes should be gated and equipped with auxiliary primers. Auxiliary primers allow the intake to be primed while delivering water from the onboard supply, thus speeding transition time.

Fig. 6–2. Know the length of hose available to quickly and efficiently establish water supply. (Illustration by Fire Medic Art)

Tankers required to aid in initial supply and attack must be suitably equipped to handle the dual role as requested. Careful thought should be given to ensure apparatus design and equipment; consider the primary role as the top priority. Tankers are well suited to be the storage compartment for tools and appliances unique to areas of rural or limited water supply. Large or expensive appliances, which do not need to be carried on individual pumpers, can be left on the tanker to save space and money department-wide. Of course, we must be sure that the apparatus will be available when and where this equipment is needed.

The pump should be of equal size as other first due apparatus in the case that it must be used as part of the attack. Plumbing should be able to adequately supply LDH at the initial flows required. The tank should be sized considering both water needs and overall apparatus weight. Having more water is great, but if the truck is too heavy to get where it's needed, the water can't help.

Getting the water off the apparatus and onto the fire quickly will increase your overall gpm. When used in stationary service, it becomes little more than a pumper with an oversized booster tank. This option requires a tank-to-flow rate equal to or greater than the fire flow need. The tanker will need to supply fire flow and the refill needs of other attack apparatus in its direct supply. This may require some creative engineering to achieve the needed flow from the tank when planning new designs.

If you plan to "dump and run," the truck should be able to quickly deploy and fill drop tanks (fig. 6–3). Standard placement of portable tanks are typically within simple reach of average sized firefighters, such as mounted upright in compartments built into the internal truck tank or on external automatic tank racks. The storage option should allow for two firefighters to be able to remove and deploy the tank.

Fig. 6–3. Drop tanks should be available for quick deployment. (Photo by Steven Heidbreder)

Hard suction and low-level strainers should be stored together on the apparatus within reach of a firefighter standing on the ground. Some departments have gone to short preconnected suction and strainers on their supply pumpers to speed this type of setup. The operator sets the tank and then swivels the assembly into the tank.

Off-loading should be another speed based consideration. The faster the portable tank can be filled to a useable level, the less likely there will be a lapse in water supply. Larger, unrestricted dumps should be the standard. Adding a plumbed jet siphon to the dump can greatly reduce dump times. For instance, a 3,000 gallon tanker can gravity dump its load in just under 3 minutes, or under 1 minute if a jet assisted dump is engaged.

Water Supply Systems

Individuals responsible for supplying water on scene should be required to have a good working knowledge of the local water system. System maps, conveniently located on response apparatus, can help with this challenge by allowing fire agencies to color code or otherwise denote hydrant locations and details. Having such information at our fingertips can allow us to make choices on which hydrant to use next or where tankers can refill without unduly restricting fire flow at the scene.

Limited supply systems

As many areas grow, the water mains are sized and placed near the intended customers. These water lines essentially become long, dead-end lines as they are commonly single direction offshoots from the supply source, or from the end of an existing main. The system may be intended to expand and take on the look and feel of its big city brethren, but funding will first be required from an adequate number of new customers. Similar to fire agencies, water supply districts cannot operate at a deficit and sometimes cannot afford needed updates. Potable drinking water will always be the water supplier's necessary and primary concern.

Sometimes water mains are adequate for the routine room and contents fires, but lack the ability to sustain pressure and flow for a major attack. If more supply is needed, consider securing an adjoining hydrant. On many suburban systems, however, this will only steal water being used by the existing supply line.

Storage and system pressure stabilizing tanks or standpipes might be used to advantage on such mains. When cost per mile or terrain limit the possibility of cross-connecting mains, the gap may be closed by utilizing strategically placed standpipe tanks that are calculated in size for a decided flow need. The standpipe should be placed as near the end of the new system as possible. This will allow for a limited two-direction supply to most hydrants on the new service. The other benefit to an expanding system is that the tanks are often reusable, and may be moved to new areas as the water system grows. Will this actually work? Only working with a knowledgeable water supply engineer and discussing your needs will tell for sure.

For a real-life example, consider a small supply system without its own well that purchased water from a neighboring supply system. The largest main in the system was the 8 in. trunk. On the far end from the supply connection, the district built an 80,000 gallon standpipe. As the water pressure dropped at the standpipe, a remote control valve opened at the connection to the other system. This valve could also be manually opened as needed. With the tank full and valve closed, volume on the 8 in. main was about 600 gpm while maintaining a residual 20 psi. Activating the supply valve increased the flow to over 1,100 gpm. Such a flow could only be maintained for about 45 minutes, but this was considered sufficient until other means of supply could be established. As a note, this small system has since been absorbed by its neighbor supply district, which saw no need for the standpipe and removed it. The flow in that area now tops out at 750 gpm.

What are the options if the volume is inadequate? Consider complementing the existing hydrant supply with static and/or mobile supply. It is important to note here that it is a public health concern to establish a water supply that could possibly backfeed untreated water into a potable water system. Pressurized water from a secondary source (including tankers filled from remote hydrants) should never be introduced to a pumper actively being supplied from a hydrant. Even with backflow protection, contamination can cause health concerns, expensive decontamination, and bad PR. Options for this are discussed later.

Limited access volume systems

What if supply system volume is more than adequate, but hydrant spacing does not allow for secondary supplies without the use of long relay lines? There are options to obtain the needed flow from a single connection point.

One solution is to use a dual pumping configuration off the hydrant (fig. 6–4). This uncommon simultaneous use of a single hydrant by two pumpers can work well where adequate system volume exists. This is a good choice any time a single pumper's plumbing or capacity is not adequate to supply the entire flow needed.

Fig. 6–4. Typical dual pumping arrangement using gated intakes. (Photo by Byron Long)

Each pumper should have at least one gated, unused LDH intake. The initial pumper attaches to the hydrant to begin initial incident supply, but leaves a gated intake accessible by the next supply pumper. When the second pumper arrives, both are hooked together, intake to intake. The second pumper now has access to that volume that is underutilized by the first pumper.

Where hydrants are equipped with both 2½ in. and LDH connections, hydrant valves can be used on the smaller outlets to allow a connection by a later arriving pumper. As with dual pumping, the first pumper connects LDH to the hydrant, except in this case also installs a gate valve on one or more of the 2½ in. hydrant discharges. The first pumper can begin uninterrupted flowing while the gate valves will allow the second pumper to connect. The downside of dual pumping is that the different connection points will affect the abilities of each pumper differently. Also, this arrangement makes it difficult to anticipate the effect from adding additional flows and the implications to the existing load.

Taking full advantage of a hydrant's volume by using multiple pumpers should not be confused with the concept of establishing a secondary water supply. The purpose for either of these flow options is to obtain the gpm needed for fire attack. If a secondary supply is required for safety, it must come from a different supply source. In limited water areas, secondary supply may require the use of nearby static supplies or shuttle operations.

Hydrants

National standards for hydrant coding are in place and used by many water supply agencies. But whose standard are they using? We are familiar with the gpm bonnet coding of NFPA, yet the same color codes are also used by water supply agencies to determine connecting main size. Thus, a flow test of a high-pressure 8 in. main may dictate a green bonnet by NFPA, yet require an orange bonnet by water supply diameter standards. If a hydrant gpm color code system is to be used on single source mains, it must be regularly monitored and updated as water demands increase.

Another hydrant marking consideration is the ability to locate the hydrant itself. Every apparatus should have a good reference map noting the location of all water sources within a given area. But even with the state-of-the-art GPS tracking, hydrants can still be difficult to spot under less than perfect conditions. Here are several options for improving the visibility of hydrants:

- Reflective street markers
- Detachable, reflective hydrant flags
- Reflective striping applied to hydrants

The active word here is *reflective*. Whatever the choice, it must be visible over a wide variety of conditions. Before settling on any one choice, experiment to determine what works best for your area. And check with local highway and private street owners as restrictions may apply.

Hydrant discharge configuration varies by area, need, and somewhat to traditional styling of the area (fig. 6–5). Occasionally, if water volume is limited, supply districts will install hydrants with only 2½ in. connection points. LDH or steamer connections are left off in an attempt to limit the flow at those particular points of the system. On many high-capacity systems—with volumes beyond the needs of any one pumper—you might find hydrants with only one LDH connection. Many hydrant connection ports can be upgraded in place, but at a cost.

When the only ports available are smaller than required, we may ignore these alternate sources or revert to smaller supply lines. A single 2½ in. discharge at 25 psi may deliver over 900 gpm, but LDH is a prime candidate for these limited connections because of its propensity to deliver higher flows with lower friction loss.

In the 1980s, a major manufacturer of LDH tried to prove the advantage of larger supply lines by using this 2½ in. port technique. An engine established a forward lay from a 2½ in. ported hydrant using a single 2½ in.

supply line. They then noted the gpm available to a master stream attached to the engine. The test was repeated using a 4 in. LDH attached to the same 2½ in. hydrant port, and the flow to the master stream was nearly tripled. Point to be taken: smaller connections can sometimes provide more flow than expected and that LDH has a definite use in suburban and rural supply.

Fig. 6–5. Hydrant discharge configuration can vary greatly. Never judge the output of a hydrant based on its configuration, color code, or static operational pressure. Accurate flow testing is the key. (Photo by 5280 Fire)

Two primary ways to connect LDH to a single 2½ in. port hydrant are by a 2½ in.-to-LDH adapter or using short, 3 in. "pony" lines siamesed to an LDH connection. Using both ports to supply the LDH may seem as though more water could be supplied, but the configuration of the hydrant ports and/or the siamese device may likely create turbulence that could actually decrease the flow. Experiment with the hydrants in your area to find the most efficient way to get the water out. Whichever is decided, always place an engine on the hydrant.

An engine placed on the hydrant can help overcome any minor losses between the hydrant and the supply line. We must consider friction loss in the supply line, even with LDH. At 1,000 gpm, a 5 in. LDH has 8 psi of friction loss per 100 ft; at 500 ft we've already lost 40 psi. On low-pressure systems, this loss may prevent us from obtaining the volume we need or require the use of a booster pumper at the hydrant. Always use the largest intake ports to connect the pumper to the hydrant, even when using short, 3 in. "pony" intake lines. Note, though, the friction loss in these small hoses and the restricted plumbing on auxiliary intakes may sacrifice the booster pump advantage.

Static Impounded Sources

In-ground static water supplies, natural or manmade, frequently offer challenges not encountered with other sources. We usually ignore or bypass such resources when we can, but with a little planning and possibly a few new devices, these supplies can be a valuable resource year round.

The best time to determine the location and viability of a static source is before the fact. Web-based search programs and mapping can give overhead satellite views of ground water sources, and some will also provide a three-dimensional ground perspective and size estimate of a given site. Never rely solely on these descriptions, though, because what looks good from a bird's-eye view may not be quite the same at ground level. To be considered a useable source, it must contain adequate volume and access.

Using the approximate surface size of the impoundment and the calculation of a cubic foot of water is approximately 7.5 gallons, we can determine the volume of that first foot below the surface. In general, assume that a body of water with a surface area of a half acre (that's less than half a football field) contains over 163,000 gallons in that first foot.

Why only the first foot below the surface? Except in extremely clear water, it will be difficult to determine how the dimensions of the impoundment change more than one or two feet below the surface. Also, dropping the level over one foot may cause environmental concerns such as runoff erosion or habitat endangerment to fish, wildlife, or livestock.

Another reason to only consider only the first foot or so of depth is due to the drafting ability of your apparatus. When using a standard suction hose and strainer, you must maintain at least 1½ ft to 2 ft of open clearance around the strainer to prevent eddies or intake of bottom debris. To maintain the rated performance of the pump, we can have no more than 20 ft of length and 10 ft of lift. Even with the apparatus parked directly beside the impoundment, hose limitations usually will not allow dropping the water level much beyond the first foot.

Once determined to be a useable option, the site should be given a unique identifier and added to mapping or water supply preplans. Information on the site should include at least the following:

- How to access the site from a known location
- The approximate amount of useable water
- Any specialized equipment that would be needed

Flowing Streams

What about flowing streams? A simple, common formula for the volume of a stream of water is to estimate the average width in feet and multiply by depth, then multiply by 7.5 for the volume of a 1 ft cross-section of the stream. To figure the flow per minute, multiply the volume by the current's speed in feet per minute. To do this, measure the distance an object floating mid-channel travels in a given time. For instance, a stream with average width of 4 ft, a depth of 4 ft, and a flow rate of 10 ft per minute equals around 1,200 gpm. Depending on the type of drafting equipment used, a sizeable amount of this flow may become a usable advantage.

We still must remember the environmental and access concerns noted earlier, along with an additional access issue. Flowing streams are continually changing their shapes as water levels change. Without solid access points, an embankment that is accessible today may be impossible to reach after the next hard rain. Access to flowing streams should be limited to points where reliable access remains stable over time.

Of course, volume on any static source requires that we can get close enough to use it. On new construction where impoundments or water runoff retention is required, encourage or mandate that an all-weather surface sufficient to allow apparatus access is installed. Don't forget to consider such requirements, even where hydrant service exists, as it can serve as a reserve supply.

Bringing the Static Source Within Reach

If an all-weather road cannot be used, bring the water to the road via a dry hydrant. Dry hydrant construction should keep overall lifts below 10 ft; a 6 in. connection and an 8 in. pipe with 45 degree elbows should be used. A dry hydrant 250 ft from the source may supply up to 1,200 gpm.

As an alternate to an in-ground dry hydrant, use a Venturi eductor device (fig. 6–6). Such larger flow devices, when coupled with a 5 in. LDH, can supply 2–3.5 times more water than required to operate the siphon. General rules that apply to these devices would be to limit lifts to less than 10 ft and the water source should not be more than 250 ft away. These siphons can also be an advantage to assist with drafting, especially when extended lift or length may be required. In a recent test, a local department used the Venturi eductor as an assist device to deliver over 750 gpm through 50 ft of 6 in. hard suction with a 28 ft lift near sea level.

Fig. 6–6. Venturi eductor prepared to deploy to a static water source. (Photo by Byron Long)

Moving Water from Remote Sites

Tanker shuttle

The tanker shuttle is usually the first choice where water supply is limited due to its relative ease and speed of setup. It works well for most initial attack considerations, but can fall short where heavy, long duration flows are concerned. The shuttle can be used in conjunction with other sources to create a combined flow to meet the needs of larger incidents or to maintain initial flow while other sources are established.

For the tanker shuttle to be effective, it must have adequate resources for the flow to continue without interruption. So, determine the rolling gpm of any tanker anticipated to respond. To simplify this process, pick the point in your area with the longest distance from a quality supply source. Have a full tanker drive the route to and from, following normal traffic rules, to establish a travel time. Now assess your time requirements by using your supply source to dump and refill the tanker. Where the tanker fleet varies greatly in size, it is advantageous to group the trucks by general size for the tests.

Consider these grouping ranges:

- Group A: 1,000 to less than 2,000 gallons
- Group B: 2,000 to less than 3,500 gallons
- Group C: 3,500 to 6,500 gallons

Also group the majority of apparatus into single axle (group A), tandem axle (B), and tractor drawn (C) units. Although the gallons carried may differ in each range, the general characteristics of each group should be similar.

Next, determine the available water in the tankers. International Organization for Standardization (ISO) testing usually subtracts 10% of the total volume, but in real life this may still be unrealistic. Better to subtract 20% to account for unlevel dump or fill sites, spillage, and incomplete fills or dumps.

Now, divide your total time (dump and fill + round trip travel) by your tank volume less 20%. For instance, a 3,000 gallon tanker with a total time of 15 minutes divided by its recalculated volume (2,400 gallons) develops 160 gpm of usable fire flow. Knowing this worst case scenario and the fire flow required, a water supply manager can now determine whether or not available resources will be adequate. Of course this system can be made more precise and thereby more effective, but by calculating worst case you can overcome many unknown or unforeseen obstacles before you encounter them!

Portable Tanks

Where the situation dictates the need to keep tankers mobile or higher flows than a nurse operation can sustain, portable tanks become essential. Where increased flow is the need, multiple tanks are often required. There are a multitude of opinions on the best way to lay out the tanks for performance, and all have advantages under the correct conditions. The final layout depends on your needs. Primary goals for a good site should be the following:

- Level, flat tank deployment area
- Easy access for the drafting apparatus
- Multiple dump points for shuttling tankers
- Ample maneuvering room for tankers

There are other water supply uses for the venerable tank than just a place for tankers to drop water. Mentioned earlier was that poor hydrants can be supplemented by a static supply. Combining potable and non-potable in the pump may lead to water system contamination, but a solution is to set a drop tank (fig. 6–7).

Fig. 6–7. The supply pumper can draft off the drop tank as tankers supplement the supply as needed. (Photo by Steven Heidreder)

Simply set up drop tank operations near the supply hydrant. One or more gate-controlled lines can be attached between the tank and hydrant to act as initial supply. It is important that these lines have sturdy anchor points on the tank and have a means to control line pressure in addition to the hydrant stem. The supply pumper can draft off the drop tank as tankers supplement the supply by refilling from sources that would not interfere with the primary hydrant supply.

If adequate volume is an issue at the fire scene, it is likely there could also be an issue at the fill point. It is always a good idea to have an engine crew at the fill site to assist tankers with connections. If that site has poor volume, why not use the pumper to advantage? As with the last example, consider using lines ported from a hydrant to a drop tank to allow the drafting pumper to refill tankers. Fill times should be reduced if the hydrant is continuously filling a portable tank equal in volume to the largest tanker. This should create better overall performance of the shuttle.

Relay Lines

When choosing relay lines as the supply option, always ensure that adequate initial water supply is in place prior to installing the relay. Thought should be given to adding at least one tanker to any area, even where hydrants exists, when a relay supply may be needed. The tanker will help to sustain initial fire flow until the relay is up and running. With LDH on the initial tanker, the tanker itself can help to install the relay (fig. 6–8). Five hundred feet of 5 in. can be stored in a tray 10 in. wide × 30 in. tall × 14 ft long, which is the equivalent of the running board alongside an elliptical tank.

Fig. 6–8. With LDH on the initial tanker, the tanker itself can help to install the relay. (Photo by 5280 Fire)

Create relay apparatus staging points that allow critical equipment to enter before the line hits the ground. Allowing late arriving equipment to drive on or around your supply line is a big safety concern.

Relay lines can be used effectively in both hydrant and non-hydrant areas. The key is to preplan and train personnel on the required skills. Some may argue that relays require too much time to set up. At 5 miles per hour, it takes less than 3 minutes to deploy 1,000 ft of supply line.

When we discuss relays, we are talking about a minimum of dual 3 in. supply lines. Even at 500 gpm, friction loss in a single 3 in. line that is 1,000 ft long puts the supply engine discharge pressure in excess of 200 psi. A better consideration is to switch to LDH. Using 4 in. line in the same circumstance creates an engine pressure of approximately 70 psi of friction loss, and with 5 in. you are at 40 psi.

A common concern on LDH relays is that too much water is used just to fill the lines. With the largest commonly used LDH, 5 in., we are only talking about 1 gallon per foot. One 2,500 gallon tanker can fill almost a half mile of 5 in. supply line by itself. Often, the apparatus that lays the LDH line has a full water tank that would normally go unused. Consider using this water to begin filling the supply line, and refill the apparatus tank after the relay has been established. Alternately, add additional tankers on the response to overcome the initial shortfall.

Another concern is that LDH is too difficult to pick up after use. Most LDH is coupled as 100 ft. sections as opposed to the standard 50 ft. So, every 1,000 ft of line is 10 sections, usually reloaded as a team effort, instead of 40 sections of dual 3 in. supply line typically rolled and handled individually. Most LDH is rinsed and reloaded on scene, while 3 in. usually requires a trip through the washer with dry time before reloading.

Whatever the size, supply hose loads should be standardized to simplify relay operations. Where required flows would not exceed 800 gpm over fairly level terrain, 1,000 ft of 4 in. or 5 in. LDH can be adequately supplied by one engine. This will still leave sufficient room in most hose beds for additional medium diameter attack line. Over 800 gpm, friction losses start to increase quickly; even LDH supply line loads may need to be reevaluated and downsized.

Relay setup

When an incident involves a structure far from the main road or water source, how do we determine the number of trucks needed?

First we need to consider the following:

- What is the required flow?
- What is the distance from fire to supply?

- Is there an elevation change over 50 ft?
- What is the size of the supply line(s)?

Considering we would not want to operate pumpers over their rated capacity for long periods, we would need to limit the friction loss to 130 psi between pumps. So, if we consider the apparatus carries 1,000 ft of supply, this limits us to the following gpm:

- 1,250 gpm for 5 in.
- 800 gpm for 4 in.
- 400 gpm for 3 in.

Relay options must also account for elevation change. Where supply lines are short, rarely do we have a need to consider this, but remember that every 50 ft in elevation changes our pressure by 25 psi and thus longer relays can be affected greatly.

An alternative to constantly calculating each relay would be to complete a water delivery overview of your entire service area and set a maximum required flow rate. With this in hand as well as knowing the limitations of your supply line and the local terrain, you can create a standard supply line load. The downside of such a system is that it will be equipment intensive on smaller incidents and will require regular reviews of your fire flow needs.

To speed any relay operation, apparatus should have the ability to initiate their portion of the relay in a forward motion. This requires some sort of marking system. The system can be as simple as setting cones or temporary markers (these markers can be spaced using a vehicle odometer at every tenth of a mile (528 ft) or as complex as creating detailed preplan maps indicating connection drop points. Another option would be to create a fixed relay marker system.

A marking system should use a standardized color and reflective device sturdy enough to act as an anchor point for the supply line. Each marker should have a unique identifier to speed relay operations. Markers should be spaced 100 ft shorter than the supply line load to account for pump connections. An apparatus assigned to the relay is given a marker ID to start its supply deployment. Knowledge of fire flow needs and the limitations of supply hose and pumping apparatus can be combined to customize the marking system for maximum performance.

Relay deployment

Each engine in the relay deploys its section and connects in, even if initial fire flows do not require the pump. By connecting in, the engine can help to fill the relay lines, act as an inline relief valve or air bleed point, and increase pressure should the need for more flow be required.

Rural Hitch Relay Connection

A means of channeling water will be required wherever tankers will nurse/shuttle water through the relay line or where water supply to the relay line will transition. A *rural hitch* combines a gated or clappered wye connected to the end of the relay and 2 in., 2½ in., or 3 in. fill lines connected to the opposite end of the wye (fig. 6–9). A tanker positions near and connects to the wye using only one of the open fill lines. The tanker operator calculates the engine pressure and begins supplying the relay until near empty. When another tanker arrives, it will connect to the remaining open fill line and charge its fill line to the same pressure as the first tanker. The first tanker is now free to disconnect and refill. Assuming more full tankers are available, a third tanker can take position at the fill line used by the first tanker and prepare to take over from the second tanker as it nears empty.

Fig. 6–9. A rural hitch arrangement at the supply end of a relay line. Here a clappered wye is used to eliminate the need for someone to switch between 3 in. fill lines as incoming tankers switch off with those in place. (Photo by Byron Long)

This cycle can continue, with each new or returning tanker taking position at the unused fill line and waiting for further instruction. As long as there is adequate room to maneuver tankers in and out of the site and valves are correctly turned on and off, steady flow can be maintained.

Although this style of supply is generally used for flows under 500 gpm, it can be used to advantage for higher flow "blitz" style attacks. Two tankers simultaneously pumping into the hitch's wye can supply 1,000 gpm for several minutes. This requires operators on both tankers to coordinate closely so that both tankers empty equally. The attack must be monitored and shut down at the attack pumper before either tanker is empty in order to prevent equipment damage. As a safety measure, time the attack to only use 75% of the total available tanker water.

The hitch can be also converted to other supply options without being shut down. A common choice is to convert to a drop tank operation to handle higher flows. Once portable tanks are in place and ready to be filled, a pumper is positioned to draft and connects to one of the fill lines. Once water is flowing, the second fill line can be removed from tanker use for use by the draft pumper. All tankers are then used for maintaining the water in the portable tank. This also helps to maintain a steady pressure on the supply line.

Where the hitch becomes a connection for a newly placed relay supply, the process is similar. A relay pumper is placed at the wye for connection to a fill line and the new relay for intake. Once air is bled from the relay, water from the new source can be directed to the fill line. The tankers are removed from the operation and the second fill line is placed in service by the relay pumper.

Common Static Flow Equipment

When considering static water supplies, a review of necessary equipment is in order. Lightweight and flexible, hard suction hose is easier to work with and inspect than its more rigid rubber counterpart. Increasing the section length from the standard 10 ft to 12 or 15 ft can allow us to use only one section on certain applications. Having fewer connections means fewer leaks, and when coupled with the lightweight hose, means faster deployment. There can be some gpm loss when adding length to standard suction lines, but this can be offset by using larger or multiple intakes. Where longer suction would be required, specifying larger pumps can assist in maintaining a desired flow.

Consideration for proper strainers is needed when suction hose is required. Just as we carry different nozzles for certain applications, no one strainer works well in all situations. Specify your needs and then test the appliance before you purchase.

Based on design and use, low-level strainer flows can vary from less than 1,000 gpm to over 1,500 gpm. Some can also double as jet-siphon devices. A siphon equipped strainer can have other uses, such as the following:

- Tank to tank water transfer
- Emergency priming aid
- Closed circuit recirculation line

When used as a transfer device, the strainer is placed in the source tank. The strainer works as venturi to use a small amount of water from a 1½ in. or 1¾ in. line to force a larger flow through the strainer and into another tank. To operate the siphon/strainer as a primer we must have three things in place:

- A gated inlet with external air bleeder
- The onboard booster tank at least half full or an outside pressurized water source
- The pump full of water

Connect the strainer and suction line as normal before connecting the strainer to a 1½ in. or 1¾ in. siphon supply line. With the inlet closed and the external bleeder open, charge the siphon supply line. When air is expelled from the bleeder, close the bleeder and open the inlet. If the supply line was being fed from the pump being primed, slowly close off the tank-to-pump valve. If the pump pressure remains steady, prime has been achieved; a pressure drop indicates an incomplete prime. In such a case, repeat the process.

To operate the strainer as a closed loop recirculation line, you need only connect and open a supply to the siphon. This works just as the onboard recirculation line routes water from the pump back to the booster tank. In this case, instead of going to the booster tank, water is routed to the draft tank. Since water moves from tank to pump and back to tank, no water is lost. An added advantage to this is with the siphon supply line charged a slight positive pressure is always on the suction line, thus greatly reducing the chances of even a weak pump losing its prime.

With adequate pressure, a 6 in. siphon/strainer can be expected to discharge 400–600 gpm in this manner. A common oversight when using these devices in multiple is that it takes valuable pump gpm to operate them. Where large flow between tanks is required, it is wise to dedicate one pumper to the siphons and another to operate the supply line(s) to the fire.

These last options are some suggestions for using low-level strainers for some common drafting operations. But they are designed for, and best utilized in, clean water drop tank applications.

If drafting will occur from natural sources, a floating strainer could be considered. Some of the same concerns apply here. To save on weight and to limit size, some models may not achieve the needed gpm.

Of course there is always the reliable barrel strainer. Properly sized and within limits, it will deliver the rated capacity of the pump. The disadvantage of the barrel strainer is that, unlike the other devices mentioned, the barrel must maintain clear water for 2 ft in all directions; otherwise, foreign debris or whirlpools may inhibit water flow.

Where static lifts may cause problems in achieving or maintaining a prime, the addition of an inline foot valve diaphragm to the strainer may help. With a foot valve, the suction line can be precharged from booster tank water and, once prime is achieved, the diaphragm will maintain the water column. These valves can be very effective when they are maintained.

A few words of caution when using foot valves. Breaking down draft lines with foot valves installed will require extra personnel as there is approximately 10 pounds of water in every foot of 6 in. hard suction.

Portable and Floating Pumps

Volume-type portable and floating pumps can be of great use. Such pumps should have a reliable, simple means of priming and/or incorporate a foot valve strainer to achieve and maintain prime under adverse conditions. Of course, these can be large and cumbersome, a problem which causes them to be left back in a corner of the station with rare use or inspection. The purchase of a small trailer easily connected to a staff SUV vehicle can solve the storage problem, and in many situations allow transport directly to the site of need.

Another alternative is to investigate the newer models of pumps commonly known as *trash pumps*, which can pump water along with small solids. Some of the smaller 2 in. pumps, delivering more than 150 gpm, are capable of being handled by one person and can self-prime. The downside is the pump connections are pipe thread and must be adapted for fire service use. Although they can produce useable flows, it is usually at pressures less than 50 psi. This does not eliminate them from consideration, as they can complement drop tank fill sites and, when used in multiples, work as fill points for tankers.

Conclusion

Water is, and will likely remain, the primary extinguishing agent well into the future. Without adequate water applied in correct proportions at the appropriate time, we are severely limited in providing effective life safety and property conservation that a given situation may require. Understanding the limitations of our current water supply will allow us to adapt responses to maintain adequate flows when needed.

Chapter 7

Special Considerations: Commercial Structures

Commercial Structure Firefighting

Fighting fires in commercial structures should be approached with different tactics and strategic approaches than fighting fires in single and double residential occupancies. These types of fires will require more resources and present with extra challenges. Suburban areas, regardless of initial intent, have become highly populated with commercial structures of different occupancies. If you use the same game plan that you have established for residential homes, you may be missing a number of variables that could lead to operational failure (fig. 7–1).

The main problem we encounter as firefighters, especially in suburbia, is complacency. Many of us fight the same fire over and over; however, we know that this is not an accurate statement. Every fire is unique. We often miss the uniqueness of each incident and are lucky enough to get away with this, flirting with disaster. This is especially true for commercial fires. Residential fires occur at a much higher frequency than that of commercial fires. For many of us, the event of the commercial structure fire is a rarity and may only happen once or twice every year. Therefore, we rely on the same tactics for a house fire that we would for a large box store or strip mall. Unfortunately, this gamble is waiting for failure and utter disaster.

When suburban developments were first established, they were intended with the purpose of being the residential neighborhoods for the urban areas. Large cities were the main areas to find commercial type occupancies. However, as the suburbs grew there was a need for commercial centers to meet the everyday needs of all the residents. Strip malls began popping up with large box stores soon following. Now, because of cost incentives and special taxing districts, even larger commercial processes, including industry, can be found in suburbs. But were suburban public services ready to handle this phenomenon and have they accounted for this boom?

You don't need all that gear...
it's just a small fire!

Fig. 7–1. The main problem we encounter as firefighters, especially in suburbia, is complacency. (Illustration by Fire Medic Art)

Sure, many of us deal with substandard staffing and less resources than our urban counterparts; however, we still can accommodate services to protect these occupancies. Plans can be established that will provide efficient tactics and strategies to attack fires in larger commercial structures.

Much of this begins with preplanning and being familiar with the buildings and understanding their physical properties. The second component is creating suggested operating guidelines (SOGs) to accommodate your agency's circumstances/capabilities and implemented to efficiently achieve operational objectives.

Examples of Commercial Structures—Type and Occupancy

Commercial firefighting is a real challenge that suburban fire agencies throughout the country face. No longer are we considered just bedroom communities. We must be ready for these challenges and have tactics to combat fires in all different types of structures and occupancies. Some typical examples of these structure and occupancies include the following:

- Strip malls
- Schools
- Indoor malls
- Apartment buildings
- Recreational areas
- Nursing homes
- Hotel/motels
- Industrial areas
- Churches

Extra Challenges

Commercial fires require different tactics due to additional challenges associated with building characteristics, occupancy types, and special circumstances. What characteristics present special challenges for commercial structures in relation to tactics and strategies? Account for these and other challenges when planning tactics for commercial incidents:

- Access (due to street access and security issues)
- Construction features (including large open areas, long spans of structural members, hidden compartments of concealed fires)
- Extended hose stretches
- Above grade floors
- Occupancy (multiple victims, invalids, children, etc.)
- Fire load
- Search and rescue challenges
- Ventilation
- Utilities
- Special circumstances/hazardous materials

Access

Often, we take structure access for granted. Commercial structures may be inaccessible due to many factors. These factors include both physical and security features. Physical inaccessibility factors can include

- Large setback from the street level
- Traffic (fig. 7–2)
- Landscaping features
- Structures in close proximity

Access considerations should not be reserved for just entrance/egress locations, but also for ladder access. Is it possible to get to floors above grade for rescue considerations and for the placement of elevated master streams? Often, because of setbacks and parked vehicles, this can become a challenge. Also, it may not be possible to have full aerial access to all five sides (including the roof) due to other structures built in close proximity. For example, in my immediate area there is a five-story hotel where there is a complete obstruction to all aerial devices on the C side due to a parking garage that does not support the weight of any fire department apparatus (fig. 7–3).

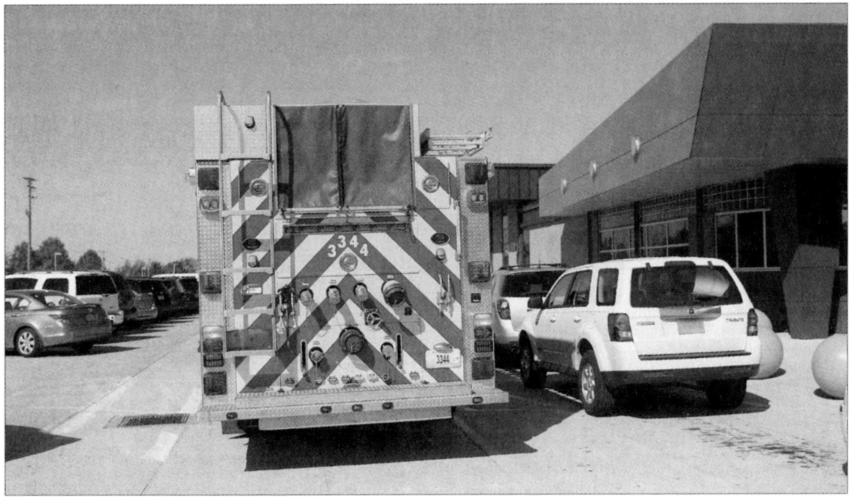

Fig. 7–2. Because of setbacks and parked vehicles, commercial structure access can be a challenge. (Photo by Erin McGruder)

Fig. 7–3. A five-story hotel with a complete side C obstruction due to parking garage placement. (Photo by Erin McGruder)

Because commercial structures may house large amounts of highly valuable content but are often unoccupied after business hours, extra security features may be incorporated at all points of entrance/egress. These structures might be equipped with a lock box device that contains keys to the structure, intended for use in case of emergency. Even if your fire apparatus has a system or key to open the box, do not assume that every commercial structure has this feature and that the keys in the box are always up to date! Fire department personnel should be familiar with doors and locking mechanisms through preplanning and have solutions to effectively gain access.

Construction features

Fires in commercial structures react and burn differently than those in residential structures due to construction features and characteristics. While it is very true that commercial structures often have a higher fire rating and class due to code, do not let this fool you. Nothing is absolutely fireproof. In fact, fires in commercial structures have the potential to burn faster due to longer truss spans and larger open areas, allowing unimpeded fire spread. These structures have many hidden spaces that can allow a fire to hide for longer durations without detection, aiding in fire growth and development.

The large open areas and taller ceiling heights can also mask the intensity and volume of fire. A fire in a residential structure will quickly consume its room or container, allowing the heat and smoke to overflow and travel away. This is not the case in a commercial structure. Envision a storage rack on fire in a warehouse-type retail store. How long will it take a fire to fill up the large box structure and begin to push heat and smoke from its entrances? The answer is a very long time. Often these fires will present with a moderate to light smoke showing, if that.

Hose stretches

Hose stretches will present a greater challenge and require more resources in commercial structures. Commercial structures have many more obstacles and stretches tend to be longer than the average residential variety. It is very obvious that stretching in commercial buildings can be a challenge and will require different tactics and more manpower. This is not a new concept and it is not even a suburban concept. In fact, the FDNY operating procedure details two engine companies for a 2½ in. commercial stretch above grade.

Hose stretches represent challenges in commercial structures due to

- The distance from the pump to the fire
- Obstacles
- Stairwells
- Floors above/below grade

In many cases, the 200 ft preconnect will not be effective for commercial structures. Often this line is way too short, especially when the fire is located numerous stories above or below grade. Also keep in mind that commercial occupancies have heavier than average fire loads, requiring larger lines. Many different agencies throughout the United States require a 2½ in. attack line for all commercial structure fires, regardless of fire volume (fig. 7–4).

Fig. 7–4. Commercial occupancies have heavier than average fire loads, often requiring larger lines. (Photo by 5280 Fire)

Consider the large commercial boxes such as the home improvement warehouse-type stores. Imagine in the middle of one of these stores, a forklift fire spreads to rows of merchandise. I am aware that these buildings are required by code to have sprinklers, but for this example let's throw in the circumstance of a system or engineering failure. What would it take to stretch to the middle of the store? How long of a stretch is possible and how much water would you need to flow? There are many "what ifs" and variables associated with this challenge; however, there are some you can be guaranteed to encounter:

- Numerous obstacles and shelves
- Numerous bends and corners
- A stretch longer than 200 ft from the apparatus pump
- Heavy fire load

Another important tactic to consider when fighting fires in commercial structures is the utilization of the auxiliary fire system (standpipes). Many commercial buildings are so large in area and height that some stretches are completely impractical. Often this is the case in structures with numerous floors above or below grade. Some single-story large box store structures are equipped with standpipe systems on the same floor as the fire, but without protected egress or area for retreat, which are contraindications to using these systems. Standpipe operations are discussed in detail later in this chapter.

Above grade floors

Access to fires in commercial buildings can be a problem due to elevation. Fires located numerous floors above grade can present tactical challenges that far exceed residential difficulties. Making access to these fires requires time and manpower. Depending on your organization, it could take minutes to ascend each floor to your destination and deploy the attack line. With this in mind, remember the time that it takes to reach the fire adds to the time the fire can continue to grow. Also remember that this can be physically demanding on firefighters and require extra resources, including extra air bottles, due to the distance from the apparatus.

Stretching lines up stairwells is not an easy task. It requires a game plan and a training component. When stretching hose up a stairwell, flake the hose wide and be careful not to get the hose caught under railings. It is also much easier to stretch hose dry; however, be very cautious when entering the fire floor with a dry line. There are questions that company officers and chief officers should answer:

- How much can one suburban crew realistically complete?
- How much can my agency accomplish with a single resource company?
- What tasks are necessary to accomplish an above grade stretch?

- How many companies will it take to make the stretch, and to maintain and support the stretch?

Occupancy type

Life hazard is the highest priority when considering tactics and strategy in relation to structure fires. It is very obvious that certain commercial occupancy types can include challenges to firefighters:

- Large numbers of occupants
- A diversity of demographics
- Challenges for mass egress
- Special needs occupants (including children and invalid assists)

These are all factors company officers should consider if the common use of the building is known. For example, when responding to a fire in a nursing home, you should automatically consider extra resources for the immediate removal and rescue of occupants, as well as request a mass casualty incident response.

Incidents that require the removal of a large number of trapped occupants require extra resources and create challenges for decision making. Upon arrival, and throughout the incident, company and chief officers should attempt to locate both the hazard and the probable location of occupants. Efforts should be made to separate the two and begin evacuation as soon as possible. Success in this undertaking is reliant on preplanning, knowing the structure in relation to its occupancy type, and making an accurate size-up. Knowing the occupancy patterns for the time of day is extremely important for decision making.

Fire load

Once again, do not get fooled by the building rating of the commercial structure. If a building is required to be rated as a Type I (fire resistive) or Type II (noncombustible) structure, that merely suggests that the building components are rated. It has nothing to do with the building contents. In commercial structures, the contents that could be consumed in fire have a much greater potential to release higher volumes of heat. Because of this fact, when company officers arrive at a commercial structure, they should immediately consider delivering a higher gallon setting and using a larger attack line.

Reference the following conversion table (table 7–1).

Table 7–1. **Higher fire loads will require more gallons per minute per 100 sq ft.**

Fire Load Type	GPM per 100 feet squared	Conversion Factor
Light fire loads (residential, offices, classrooms)	10	0.1
Ordinary hazard areas (most commercial)	20–30	.2–.3
High fire loads	30–50	.3–.5

Do not automatically assume that the residential 1¾ in. attack line will always work just as well for commercial structure fires. This tactic often fails. It is a very simple concept: water absorbs heat at specific, well-tested ratios. Commercial fires burn hotter and so require a prescribed amount of water, often delivered by a 2½ in. attack line set at flowing 250 gallons per minute.

Search and rescue challenges

Many commercial structures are large in terms of area and have vast, open spaces. When there are conditions of zero- or low-visibility, conducting a search for trapped occupants can be very difficult to say the least. Search and rescue tactics are not the same for commercial as they are for residential. If you decide to conduct an oriented search off a hoseline in the middle of big box home improvement store, it is very possible that later crews will not only be searching for the trapped occupants, but also for you and your crew!

Search and rescue in a large commercial building will require specific tactics. Chapter 9 highlights important considerations for search functions in both commercial and residential buildings and describes the different tactical approaches for each. The search and rescue function in commercial structures requires training and a systematic approach.

Ventilation

Ventilation is another fireground activity that may be performed in a different tactical fashion in commercial structures as compared to residential. In smaller, one-story commercial units, such as a strip mall stores, ventilation may not be too dissimilar. It may be as easy as taking out a back window and door and either allowing natural air currents or forced mechanical ventilation to enter the opposite side of the structure and horizontally ventilate. Even vertical ventilation in this instance isn't much different, except for the fact that the roofing material may be built up and require a different type of saw and blade. But what about the big box stores with a larger volume to pressurize? What about a fire in structure that is numerous floors above grade?

In the example of the big box store, it may be difficult to effectively remove smoke and heat from interior sections (fig. 7–5). Often, natural air currents are not enough to remove the smoke. Mechanical ventilation is an option to positively pressurize the compartment and force the smoke toward the negative pressure of the external atmosphere.

Fig. 7–5. Natural air currents are frequently insufficient to remove smoke and heat from large commercial buildings. (Photo by Erin McGruder)

This endeavor, especially in a large building, may require more than one fan. Another consideration is that these large structures often have ventilation hatches built into the roof for quick removal of smoke. Firefighters should be aware of these building components and be ready to properly and safely utilize them.

Fires in high-rise buildings also require ventilation for an effective coordinated attack. How do we remove smoke and heat from compartments that are not in proximity to the natural openings of the entrance? Horizontal ventilation with natural air flow is one possible solution; however, when breaking exterior windows, firefighters should be aware of what is below and the hazards of falling glass. Also, interior crews should be aware of wind conditions. Do not break exterior windows if wind conditions are extreme, especially in upper floors where the currents can be stronger. The wind can cause the fire to act like a blowtorch toward interior crews. Another important feature when considering ventilation in high rise buildings is interior ventilation systems. These systems can help remove smoke, but they should not be used during a fire condition. They should be controlled and turned off because of their potential to spread heat and fire throughout a building.

One beneficial tactic, often mandated by numerous fire agencies for high-rise operations, is pressurizing the stairwell. Pressuring the stairwell with mechanical ventilation helps secure and guarantee an anchor for both suppression activities and safe evacuation. This procedure should be detailed to companies responsible for facilitating functions on the initial first alarm response.

Utilities

Utilities, such as gas and electric, are more extensive in commercial structures and require crews specially detailed for utility control. Most commercial buildings will have large electrical feeds and numerous breakers located in a separate room. Also, gas meters are generally much larger and different from residential meters. It is important that companies preplan where these features are located for rapid control and safe operating procedures.

In addition to gas and electric, crews should be familiar with elevator and ventilation systems located within the structure. These should be detailed on all building preplans and schematics.

Special considerations (hazardous materials)

A final consideration fire service agencies should be aware of, through building inspections and pre-incident planning, is the presence of hazardous materials. Commercial occupancies, specifically those housing industrial processing and manufacturing operations, often require the use of chemicals and materials that can be immediately hazardous to health, unstable, explosive, or flammable. When operating in these structures, it is important to know where and how much of the materials are present in order to confine the materials and exercise caution during operations.

Commercial Structure Features and Systems

Commercial structures and occupancies have numerous features, not found in residential structures, that are extremely important to firefighting tactics and strategies. We must understand how they work and where they are located within and on the actual structure. Once again, preplanning is a critical function needed to pinpoint all their locations.

Features of interest include

- Sprinkler systems
- Standpipe systems
- On-site fire system pumps
- Large corridors and stairwells
- Ventilation systems
- Elevators
- Security systems and additional entrance/egress mechanisms
- Emergency key lock box

Equipment Needs

Commercial structures not only require different tactics, but also specific tools and equipment for firefighting. This can be attributed to the special circumstances previously mentioned and building construction features.

Commercial structure fire equipment needs include

- Special hose loads
- Hotel /high-rise packs
- Forcible entry equipment, such as through-the-lock tools and metal cutting K-12 saws
- Ventilation saws specifically for built-up, commercial roof construction
- Large area search ropes
- Personal bailout systems

Commercial hose loads

Your apparatus setup should fit its response area, especially in regard to hose loads. If your response area is detailed with numerous commercial structures and long setbacks from the apparatus positioning, then you should have an adequate amount of hose and the right load to accommodate these demands. Hose beds should not be generic, but designed as dictated by the needs of the response area, the apparatus inventory, and specific setup.

Numerous techniques exist for commercial hose loads. The right load is one your company can quickly deploy in an efficient manner. Imagine having to deploy a 300–400 ft stretch into the heart of a commercial warehouse or manufacturing plant. Does your apparatus have a hose bed lay that can be easily deployed by a company (or two) to accomplish this task? Typically, apparatus that have the potential for lays of this nature have a large static supply of reversed 2.5 in. or 3 in. hose. This load can be loaded flat, on its side, in an accordion or horseshoe. Also, these lays may have a finish load on the end that may include a bundle of 50 ft or 100 ft. This hose will be the primary line that will be maneuvered once the line is in the general proximity for placement. This finish load can be flat bundled or bundled in a horseshoe so it can be easily carried by one firefighter (fig. 7–6).

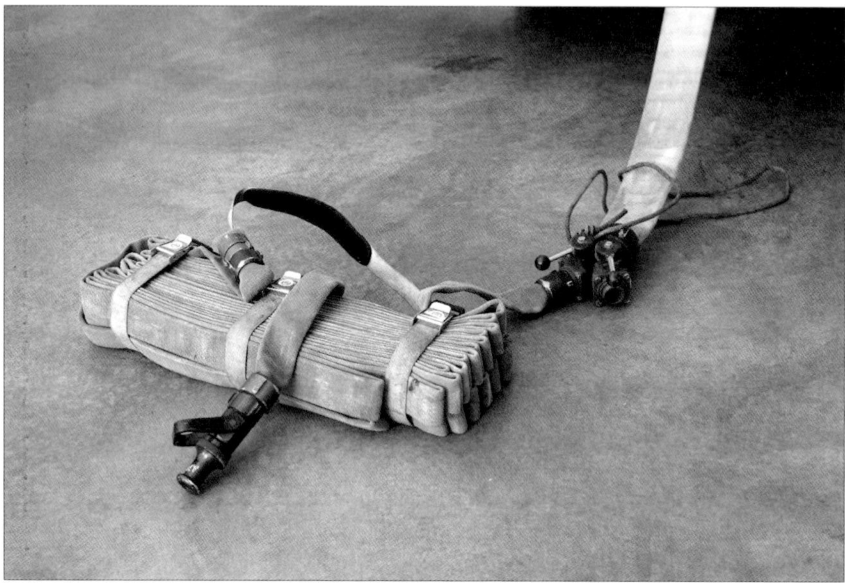

Fig. 7–6. A high-rise or standpipe pack should be bundled so it can be efficiently deployed in a timely manner. (Photo by Erin McGruder)

It was stated earlier that it is desirable to deploy at least a 2.5 in. attack line on commercial buildings; however, if your agency has a situation where you wish to deploy a 1¾ in. line over 250 ft, there are options to consider for your hose bed. One option, which my organization utilizes, is to have a 700–800 ft static supply reversed 3 in., finished with a gated wye (2½ or

2 in. × 1½ in.). If my company needs to make a stretch over 250 ft, I have the option to deploy the feeder 3 in. and have another company attach their bundled 150 ft of 1¾ in. hand line. All of our apparatus have this bundle in a compartment for this purpose and for a possible option for standpipe operations. (Note, however, that 2½ in. hand lines are desired for standpipes.)

Another option is the *minuteman* load. This load feeds from the top of the stack, but is not dragged. Each bundle is carried by a firefighter. This load is fantastic when the stretch has many obstacles in its path and a drag is far too impractical.

High-rise/standpipe kits

Another important item that should be equipped on apparatus for commercial firefighting is the high-rise/standpipe kit. Regardless of what your agency calls this setup, you should have a system in place to utilize a building's provided standpipe. This system has two components:

- The bundled hose (primary kit)
- An auxiliary bag of essential tools (secondary kit)

There are different types of hose bundles used throughout the United States for the purpose of high-rise/standpipe operations. Your agency should utilize the bundle that can be efficiently and consistently deployed in a timely manner by your operation. High-rise/standpipe bundles can be loaded with both 1¾ in. and 2½ in. hose. Often, the bundle consists of 150 ft of hose.

A bundle of 150 ft 1¾ in. hose can be easily flat loaded and held together with a handled strap. Unfortunately, this load can be bulky and heavier than desired. Also, most standpipe systems are engineered to operate hand lines at 50 psi; therefore, it is not desirable to finish the load with a fog nozzle. The smooth bore nozzle is ideal for a standpipe hose kit.

Because of fire load and the length of time required to deploy on floors above grade in commercial structures, the 2½ in. is most often the hoseline of choice. It is far too impractical to bundle the entire 150 ft of hose into one stack. Ideally, each section is bundled into a horseshoe to be easily carried up flights of stairs by a firefighter. Once again, the ideal nozzle for this load is the smooth bore nozzle. My apparatus currently uses this system; however, we divide it into two 75 ft sections. This is done because of limited personnel and the fact that the weight can be handled under the circumstances (the tallest structure in the response area is only 10 stories).

The secondary kit (auxiliary standpipe kit) should contain essential tools to assist with standpipe operations (fig. 7–7). The tools that should be included in the kit include the following:

- 2½ in. inline pressure gauge
- Miscellaneous adaptors
- Spanner wrenches
- Wire brush
- Spare or backup nozzle
- 1½ in. to 2½ in. increaser
- Short section of 3 in. hose (5 ft) attached to a gated wye (2½ in. female to 2 × 1½ in. male) (for use with 1¾ in. standpipe bundles)
- 2½ in. elbow
- 18 in. pipe wrench
- Spare operating hand wheel
- Hand wheel wrench
- Door chocks
- Door straps
- Grease pencils or chalk

Forcible entry equipment

Forcible entry for commercial structures requires different tactics and entry tools than those used for residential structures because their doors and locking mechanisms are different. Most front entrances to commercial buildings are made of glass inserted into tubular steel. They swing outward, have commercial mortise locks with large bolts, and cannot be conventionally forced (pried) with irons. The technique that should be used for gaining access through these doors is *through-the-lock* forcible entry.

Through-the-lock forcible entry requires removing the lock cylinder and manipulating the lock. The lock cylinder can be removed using specialized tools:

- The A-tool
- The K-tool
- The rex tool (a larger A shaped head with a longer prying shaft)
- Large set of pliers

Fig. 7–7. Auxiliary standpipe kits should contain essential tools to assist with standpipe operations. (Photo by Erin McGruder)

The lock then can be manipulated with a variety of keys, usually stored within the tool pouch.

In addition to the front door, there are other challenges associated with forcible entry in commercial structures. Security bars on windows and doors and extra hardware placed on rear doors, such as drop bars and dead bolts, can pose additional challenges and risks for firefighters. A cutoff saw (K-12) with a metal cutting blade is needed to remove these barriers.

Ventilation tools

Commercial roofs are often constructed of layers of built up materials, such as an under decking material covered with a tar and rock finish. The under decking may be made of either steel or wood. Unlike a residential roof, which is usually easily cut with a chain saw, cutting through the commercial roof's layers and under decking requires a saw designed for this application. One best suited for this application is the cutoff saw (K-12) with a multi-purpose blade (fig. 7–8). This blade can cut through the tar and the wood or steel under decking without gumming up.

Fig .7–8. A K-12 saw can be used to penetrate commercial roofing materials. (Photo by Steven Heidbreder)

Large area search ropes

As stated earlier, large area searches are complicated. Covering large open rooms requires a systematic approach and firefighters to be anchored (tethered) to a lifeline at all times. Searching with a hoseline for this application is simply ineffective, unsafe, and impractical. The hoseline cannot be maneuvered as rapidly as a search rope. A search rope system should be established and placed on the apparatus for this function. The length of the rope should be determined in respect to the size of the buildings in the response area. These search systems will often require numerous carabineers and other shorter lines attached to the main search rope. These systems are explained in detail in the chapter 9.

Personal bailout systems

Companies that have the potential to operate in structures with numerous floors above grade should consider furnishing their members with bailout systems. It is sometimes impossible to have aerial devices in place on every side of the structure for quick egress for firefighters, and unfortunately not all fire operations go as planned; therefore, firefighters should have an alternative for a rapid exit. Bailout systems can be built into bunker pants, mounted on SCBA, or come in a separate bags attached to bunker gear. Regardless of design, they should include the same features:

- A hook (for attachment)
- A prescribed length of small diameter rope (dependant on specific building height)
- A descending device
- Various carabineers

Today, most bunker gear is available with a built in personal harness option. If your company has the potential to operate above grade, it is advisable to pursue this extra safety feature.

Resource Needs

Commercial firefighting does present extra challenges that require added fireground activities and functions. These functions should be considered and factored into commercial suggested operating guidelines. A normal function for a residential structure fire, such as deploying an attack line, may require twice the manpower for a commercial response. This should be anticipated, and automatically requesting additional alarms should be encouraged. Regardless of the volume of fire, consider the occupancy or structure type. For example, a fire confined to a single room in a one-story residential can be easily handled with a few companies. Now place that same fire on the tenth floor of a hotel complex. That is a much different fire and will most certainly require significantly more resources.

Automatic considerations for fires in commercial structures include the following:

- Extra alarms
- Extra personnel
- Support of auxiliary suppression systems
- Additional emergency medical support for multiple victims
- Additional aerial devices
- The possible use of master stream devices for large fire volume and exposure protection
- The need for extra water sources
- The placement of engines on water sources to boost depleted system pressure

When developing SOGs for your agency, all of the factors should be considered, including these guidelines and operation details:

- Develop benchmarks for striking multiple alarms and have those benchmarks illustrated in the preferred operating document. *Second alarm assignments will be requested automatically with fire reported above grade in XYZ type commercial structures.*
- Detail an early arriving engine to support the sprinkler or standpipe system. This is *essential!* Sprinkler systems and standpipes must be supported to assist with fire containment and suppression. *The second arriving engine will connect to the FD connection and support sprinkler and standpipe systems.*
- Develop automatic responses for mass casualty incidents. *Upon arrival at a structure with numerous victims, automatically request an MCI (mass casualty incident) level 1-3. An individual MCI will detail five ambulances, an engine, and a chief officer.*
- Have additional aerial devices automatically detailed on response plans to commercial structures (if available).
- Detail a later arriving engine or second alarm engine responsible for establishing a second water supply.
- Document the mandate that if more than one water supply is established, an engine will be placed on the hydrant to supplement system pressure.

Tactics for Strip Malls and Large Box Store Construction

Strip malls and large box stores are two of the most common types of commercial structures in suburbia that give firefighters major challenges. The reason for this is the heavy fire loads these buildings represent and the lack of experience fighting these types of fires. These structures represent huge hazards for both fire spread and collapse. Sound tactics play a huge role in losing or saving these tremendously large buildings.

Strip malls

A strip mall, also known as a taxpayer, is a row of stores housed in one common structure (fig. 7–9). When they were first introduced they were known as taxpayers because they were buildings landlords could construct cheaply and that would generate rental income to pay for property taxes while anticipating a future increase in value. Older style taxpayers are generally Type III, ordinary construction, having masonry exterior walls and wood frame interior walls. Code enforcement has changed this trend and now newer constructed taxpayers are of Type II, noncombustible construction.

Fig. 7–9. Two-story taxpayer with a flat roof. (Photo by Erin McGruder)

The most important aspect of strip malls that firefighters should concentrate on is the roof construction. Older types of taxpayer/strip malls usually have one of four different types of roofs: flat, inverted, metal deck on bar joists, or bow string construction. Newer constructions usually have a roof made of corrugated metal decking laid over unprotected steel bar joists. The most dangerous to firefighters is the bow string; if fire is found in the bow string roof members, all personnel should immediately evacuate! This type of construction will not hold up long under fire conditions and will catastrophically fail. Unprotected steel joists will also fail if not cooled quickly. Efforts should be made to immediately get into the joists to check for exposure and begin the cooling process.

Another important building construction feature is the *cockloft*, the opening above the ceiling that is often used for utilities. In this construction the interior members do not extend past the ceiling, which leaves a huge space that can span the entire length of the structure. Open cocklofts were extremely prevalent before stricter building codes required fire stopping construction features; however, they can still be a major problem in the newer strip malls. The open or breached cockloft can be the culprit for spreading fire throughout an entire strip mall.

Tactics for attacking a fire in a strip mall/taxpayer entail getting ahead of the fire with a large caliber weapon, such as a 2½ in. attack line. There is the possibility of large fire load in the structures, even in Type II noncombustible. The key is to stop the spread and growth of the fire with a well coordinated fire attack. The 2½ in. should be directed into the unit that is involved, while subsequent 1¾ in. lines are directed into the adjacent units. These 1¾ in. lines should immediately go into the cockloft space to prevent any possible spread and cool steel joists. Crews should complete the coordinated fire attack with adequate vertical ventilation in order to redirect heat and smoke and to allow interior crews to keep their positions. Finally, do not forget to complete forcible entry. Crews should have two points of egress, which in strip malls means forcing the rear door.

After analyzing a few of the necessary functions involved in attacking a commercial strip mall fire, your agency can start piecing together resource needs and develop a suggested operating procedure.

Figure 7–10 is obviously an example and may not fit every operation; however, you can immediately see the minimal resources needed. It is beyond the capability for first alarm assignment for the majority of suburban agencies. The example is minimal and makes a few assumptions, such as whether a 2½ in. line can be deployed with a single engine and single facilitating company, and whether more exposure lines will be needed.

Company	Function
Engine (Quint) 1	Initial Line Deployment (2½ in.)
Engine (Quint) 2	Facilitating Initial Line
Engine (Quint) 3	Exposure Line (1¾ in.)
Engine (Quint) 4	Facilitating Exposure Line
Engine (Quint) 5	Vertical Ventilation
Engine (Quint) 6	Forcible Entry (Rear)
Engine (Quint) 7	RIT

Fig. 7–10. Sample first alarm apparatus assignment

Large box store construction

Large box stores are generally one story; however, can often have the same area as a multiple story commercial building. These large warehouse-type retail buildings are popping up all over the suburban United State and are typically Type II, noncombustible. Do not be fooled by the type of construction or the fact that the majority of these buildings are sprinklered. The fire load in these structures can be massive due to the content. Determining fire volume and location can be challenging. Size-ups can be difficult and large fires on the interior may only present with light smoke showing.

These fires will require large amounts of water. The minimal size attack line should be the 2½ inch. Do not be afraid to place a master stream in service on the interior of these structures. There is no rule stating that the master stream is only a defensive tool. Be prepared to secure numerous water sources and have aerials in place in case the response goes defensive.

Many of these structures do come equipped with standpipes. These should be avoided, as mentioned previously, due to the fact that paths of egress cannot be protected for interior crews. SOGs should indicate, however, that an early arriving engine should immediately support the sprinkler system.

This fire will most likely require numerous attack lines if the operation remains offensive. All efforts should also be made to protect any possible exposures. All facilitating functions are also applicable:

- Completing forcible entry
- Overhaul
- Vertical ventilation
- Search and rescue/evacuation
- Utility control

High-rise/Standpipe Operations

Typically, buildings are not as tall in the suburbs as they are in large cities. In fact, many of our tallest buildings do not fit the standard definition of a high-rise (fig. 7–11). For the purpose of defining suburban tactics, however, a *high-rise* is any fire resistive building that exceeds the reach of an available aerial. This definition can also be extended to any building that lacks windows, causing all firefighting activities to be conducted from the interior.

Fig. 7–11. Many buildings in suburbia do not fit the urban definition of high-rise, however must employ high-rise tactics. (Photo by Erin McGruder)

There are specific strategies that must be followed for fighting fires in high-rise structures:

1. Conduct an adequate size-up and determine the fire location.
2. Verify the fire location before committing any resources.
3. Identify and establish separate attack and evacuation stairwells.
4. Gain control of building systems.
5. Confine and extinguish the fire.

Conduct an adequate size-up and determine fire location

This function might not be as easy as it appears. If the first arriving company is lucky, fire will be showing upon arrival and the company will be able to identify the fire floor. It might also prove very difficult to get a complete 360-degree view of the structure due to its size or limited access. It would be beneficial to have other incoming apparatus approach the structure from a different direction and get the completed view.

The first arriving apparatus should enter the structure and immediately find the fire control panel, which should indicate the floor where the problem exists. Also, the officer can attempt to get as much information as possible from any fleeing occupants.

Verify the fire location

Before connecting to a standpipe system, crews should verify the location of the fire. Valuable time can be lost and this delay can make a large impact on the operation. The first officer should attempt to get a visual of the fire floor and communicate the verification to all incoming apparatus.

Identify stairwells

Once the fire floor and the location of the fire have been determined, the initial officer should establish both an attack stairwell and evacuation stairwell. High-rise operations are much more effective if these two functions are separate. It is very difficult to have a mass evacuation while also committing fire department resources in the same stairwell. The traffic is too much for an average size stairwell and a hindrance to the operation. Preferably, the attack stairwell will be in close proximity to the location of the fire, and the evacuation stairwell will be away from the fire and allow occupants to flee to a safe area, free from heat and smoke.

One very important activity that should not be neglected is pressurizing the stairwells with mechanical, positive ventilation. This will help secure the stairwells and aid in keeping smoke and heat on the fire floor.

Gain control of building systems

Crews must be assigned to gain controls of all building systems:

- Elevators must be controlled immediately and returned to the lobby. All elevators must be checked to make sure no occupants are trapped.

- All HVAC systems must be shut down and be placed in non-recirculating mode.
- If the building is equipped with a fire pump, crews must ensure that it is operating to supply the proper pressure to both standpipes and sprinkler systems.
- Gain control of the address system, if equipped, to inform and direct occupants to safety.

In order to accomplish these functions, companies must be familiar with buildings in their areas and how to operate these systems.

Confine and extinguish the fire

This can be a very resource and labor intensive operation. It obviously entails more than simply advancing toward the body of the fire with a preconnected attack line. It requires hauling an extensive amount of equipment up many flights of stairs, connecting to a standpipe system, and operating under a systematic plan to achieve a coordinated fire attack. Ventilation is challenging but necessary in order to help crews advance. In a closed structure, such as a high-rise, heat can pin down an advancing crew and prevent their progress.

An operating method my agency utilizes for suppressing fires in high-rise buildings and operating off standpipe systems is the *three crew relief* method. This method includes

- A crew for deploying the initial attack line
- A crew for assisting the initial attack crew with line deployment
- A crew staged on a floor below for immediate relief

One key element companies should be aware of is that they may not be able to accomplish both the stretch and the attack. Crews should know their limitations. It may be physically impractical for a single company to attempt to complete all the required tasks:

- Locating the fire.
- Hauling equipment up numerous flights of stairs.
- Verifying the fire location.
- Assembling a hoseline and connecting to the standpipe.
- Stretching the line onto the fire floor.
- Battling high heat conditions and flowing the attack line to contain and extinguish the fire.

The three crew relief method addresses this limitation. It allows for a secondary crew to assist with the stretch and also allows for immediate relief. Crews do not want to lose their hard-fought progress on the fire floor. The immediate relief allows for crews to keep their position and continue forward without losing ground to the fire.

There are a number of considerations for high-rise firefighting:

- The attack line should be connected to the standpipe a floor below the fire floor.
- The attack line should be 2½ in. with a smooth bore nozzle. An inline pressure gauge should also be used to assure the proper operating pressures.
- A staging area should be established two floors below the fire floor for staging extra air supply and additional crews.
- RIT teams should also be located in this staging area.
- The fire floor and floors directly above the fire should be searched.

Now apply all of this to your operation and agency needs. Can you develop an effective suggested operating procedure to account for all the necessary functions and to efficiently assign responsibilities to all incoming companies? How many alarms will you have to request in order to handle all the essential high-rise functions?

We can break down the high-rise functions into separate assignments:

- Size-up and fire location
- Initial attack line deployment
- Backup to the attack line
- Relief company
- Backup line deployment
- Water supply
- Sprinkler system/standpipe supply
- Evacuation support
- Ventilation
- Search and rescue
- Ventilation and stairwell pressurization
- Overhaul

- Elevator control
- Utility control/including HVAC
- Fire pump control
- RIT

High-rise operations cannot be accomplished with a single first alarm assignment. In fact, a third alarm assignment may not be sufficient to handle all of these functions. It is very important to analyze available resources and to realistically assess their capabilities.

Developing SOGs past the first alarm assignment can be challenging. As is the case on any structure fire, all functions must be broken down and adequately assigned for efficient and effective fireground practices. High-rise firefighting, even though more complex, is no exception.

Chapter 8

Special Considerations: Technical Rescue

This chapter was contributed by Captain Steven Heidbreder of the Metro West Fire Protection District, St. Louis County, Missouri, and member of the USAR Missouri Task Force-1.

Technical Rescue Challenge Defined

As has been discussed throughout this book, suburban fire departments are faced with many of the same challenges as their big city counterparts, often with only a fraction of the resources available to deal with them. Technical rescue incidents are just another service that the suburban fire department provides to its citizens with limited staffing, equipment, or training that is often readily available to the larger city departments. This chapter provides information that can help you when evaluating technical rescue options.

What is meant by technical rescue? Some examples include

- Water rescue
- Ice rescue
- Dive rescue
- Rope or high angle rescue
- Confined space rescue

- Trench and excavation rescue
- Building rescue
- Wilderness search and rescue/K-9 search
- Vehicle and machinery extrication

Technical rescue can be one of the most time and resource demanding services a department can provide to its community. Among other considerations, there are two important financial issues:

- Purchasing the required technical rescue equipment and personal protective gear.
- Purchasing or modifying apparatus for technical rescue equipment.

Either of these items will be very expensive. In addition to the equipment procurement costs, you must also consider the amount of time and resources needed to both attain and maintain proficiency in a wide range of technical rescue disciplines. It can take many years to develop a technical rescue plan for your department. You must be prepared for a long-term investment of planning that is in a constant state of flux. Because the cost of implementing a technical rescue plan requires long-term financial investment, you will have to be flexible in planning and implementation. You must also be willing to invest training time in your program. While the range of challenges facing the suburban department seem endless, so too are the possible solutions.

Community Survey

If your department decides there is a need to provide technical rescue services to your community, it is important to identify what types of technical rescue incidents are possible in your area. If there are no bodies of water or potential for flash flooding in your community, you won't spend much time or money establishing a dive team or swift water rescue training (fig. 8–1).

Conversely, if there is a great deal of construction occurring where trenches are dug routinely for utilities, deciding to prioritize training and equipment capital on trench rescue training would be a wise investment.

In addition to your community survey, it is important to know what resources are available from other fire departments or agencies in your town, county, or state. What are the technical rescue challenges facing your neighboring departments? Do they possess the resources or personnel to address these challenges? Is a fully staffed and equipped technical rescue team already located in a neighboring county? This may allow you to scale down your plans and budgetary needs if a mutual aid agreement can be reached with another nearby agency.

It is imperative that you conduct a comprehensive preplan of your community for its technical rescue potential and challenges. From this survey a plan can be developed and a road map laid out for your department to meet the technical rescue needs of your community.

Fig. 8–1. Consider prioritizing training and equipment capital based on the types of technical rescue incidents possible in your area. (Photo by Steven Heidbreder)

Regional Approach

Reach out to your neighboring agencies to forge agreements that address your technical rescue challenges with a multi-jurisdictional solution. Instead of a department with limited resources trying to master all the rescue disciplines needed, selective specialization might be the answer by several neighboring departments, each focusing in one particular rescue discipline.

This department will train its members to the technician or specialist level and invest their resources in obtaining the necessary rescue equipment. Members of the other departments can be trained to the operations level and used to support the department that is specialized in that discipline (fig. 8–2). By approaching your technical rescue situations with a multi-jurisdictional solution focusing on selective specialization, all your rescue needs can be met with minimal outlays of time and money.

Fig. 8–2. Low-frequency, high-acuity technical rescue possibilities, such as structural collapse, may be best maintained at the regional level. (Photo by Tom Vatterott, Jr., of Metro West Fire Protection District)

Another regional option our department was able to take advantage of was Homeland Security funding for technical rescue assets. The St. Louis metro area was able to place into service five tractor-trailer heavy rescue units spread out over its five counties. Along with the rescue assets came several rounds of training that allowed a large pool of personnel to attain a minimum of technician level certification in rope rescue, confined space rescue, building collapse, and trench collapse. This regional approach allows a tractor trailer rescue unit to respond to a multi-jurisdictional incident where it is met at the scene by crews from numerous area departments in each county.

Training

Once you have decided your department is going to provide technical rescue services, how do you get your people trained to meet the challenges specific to your area? Do you train the entire department to the technician level or do you have one company or group of personnel specialize and have the remainder of the department trained to operations level?

We found that attempting to train every member at our department to the technician level was impractical. Not only was this a cost-prohibitive proposition, but the logistics of providing technician level training to such a large number of personnel would not work for our organization. We would have also had the issue of maintaining their annual certifications with a limited training budget and limited training time outside of our normal monthly program. With these factors in mind, it was an easy decision for us to train only a core group of personnel to the technician level for our various rescue disciplines. Operations- and awareness-level training was then provided to the remaining members of the department through our normal monthly battalion training program. To address the continuing training needs we have monthly shift-level technical rescue training that covers a rotating set of skills. Once a quarter we have a team drill that pulls all members together to train on various rescue scenarios. An assessment of your department and personnel will be needed to determine what will work best for you.

When surveying your community for hazards, you can also perform a survey of your town, city, county, state, or federal training opportunities and funding. Several states offer training courses covering almost all technical rescue disciplines in conjunction with state training organizations or universities. Many of these classes are offered free of charge or for a reduced fee through training grants. A federal program administered through FEMA called the Assistance to Firefighters Grant is an excellent source of funding to bring training to your department for little money. The International Association of Fire Fighters also works with departments to secure grants for funding training programs. Another option might be to pool your resources with neighboring departments to sponsor a class that can be delivered to a core group of personnel from each department. Using this multi-jurisdictional model will appear more favorable for grants as they will cover a greater geographical region and larger population. More bang for their training grant buck, if you will.

After you have decided to establish your technical rescue team or if you already have one in place, how do you develop a comprehensive training program to keep your people and their skills sharp? A great resource for developing training guidelines is *NFPA 1670: Standard on Operations and Training for Technical Search and Rescue Incidents*, published by the National Fire Protection Association. This standard identifies and establishes levels of functional capability for safely and effectively conducting operations at technical rescue incidents (fig. 8–3). Chapter 2 of this standard even lays out a road map for you to use when planning a technical rescue team.

Fig. 8–3. Technical rescue team member training to rappel with a pick-off technique. (Photo by Steven Heidbreder)

We are also able to send several of our people to the Fire Department Instructor's Conference (FDIC) every year to keep abreast of new developments in technical rescue. This group of instructors returns and develops training programs from the knowledge gained in their classes. FDIC is one of several excellent training opportunities presented around the country annually. I cannot emphasize how important it is to get your technical rescue folks to one of these training conferences as often as possible. There are some years where five to six months worth of training material is brought back by our rescue group.

Another avenue for training opportunities is developing networks outside the fire department, such as with local utility companies, construction and demolition companies, and city planning and inspection authorities. Find out where new construction may be occurring in your area or where buildings are being razed for redevelopment. These structures can offer almost limitless opportunities for technical rescue training topics, such as breaching, breaking concrete, and exterior and interior shoring operations, to name just a few. We have been lucky enough to have a contractor collapse a house for us to use for a week of training prior to its final demolition and removal.

There also needs to be an emergency medical services (EMS) component to your training program. If your department is responsible for providing EMS, make sure to incorporate the medical aspect of treating your patients during the rescue incident. If EMS is provided by another service, get them involved. Have them give your team members classes on crush injuries, hypothermia, and other typical medical issues associated with the technical rescue discipline involved. Make sure the technicians and paramedics know what is expected of them during the rescue incident. When and where is advanced life support treatment appropriate versus basic life support (BLS)? The time to sort it all out is not at o-dark-thirty with three people buried deep in a trench collapse in the pouring rain. Work it out ahead of time and include them in your ongoing training. Remember, we are professionals. Practice and train until you can't get it wrong.

Along with stressing the importance of obtaining the proper training, gather your rescue group and any mutual aid rescue personnel together to discuss and develop standard operating procedures or guidelines (SOPs or SOGs) for your technical rescue incidents. Our region is in the process of expanding the scope of our technical rescue SOGs. No longer will the SOGs address just the technical rescue team and its operations; we are developing all-inclusive procedures that start with the first arriving fire or medical company and expand all the way up to a full blown technical rescue incident, including turning operations over to a federal urban search and rescue (USAR) team. Developing these procedures is a fraction of the challenge. If they sit on a shelf in some chief's office and never again see the light of day, you've wasted a lot of time you could have otherwise used training. These guidelines need to be second nature to all personnel. Everyone needs to be on board.

Develop an action plan to get all of your people trained on the new procedures. Begin with lectures to introduce the concepts, then get them out in the field for real life exercises to put these procedures into practice. After your initial round of training you will need to reinforce the training in subsequent months with training bulletins and drill sheets. Post them in the bathrooms and kitchens of the fire stations. Anywhere your people gather and spend time—put the training material in front of them. Begin each month's training with a brief review of the previous month's training subjects to keep drilling it into their heads. Tell them, show them, and then let them do it over and over until it becomes second nature (fig. 8–4).

Fig. 8–4. Developing procedures is a faction of the challenge. Crews must practice until technical rescue becomes second nature. (Photo by Box Alarm Production Fire Ground Photography)

Take full advantage of available technology for your training. A small investment in an inexpensive HD video camera allows us to video all of our training exercises. With a little bit of time, a laptop computer, and decent video editing software, we produce training videos for our members to review after training. You can even take it a step further and use free web hosting services like YouTube or Vimeo to post your videos online. This will allow members to review the training at a time or place of their choosing, including their smart phones or tablets!

If you have developed a multi-jurisdictional rescue team, the next step will be to conduct training with your mutual aid companies to ensure everyone is on the same page when responding to and mitigating technical rescue incidents. Get to know your neighboring departments to help foster a cohesive working environment. Technical rescue incidents can be very stressful and personnel intensive. There is no place on the scene for bickering, big egos, or grudge matches. Learn to work together as a team. Remember, we are here to serve our citizens, not our department or personal pride.

Our region was finally able to conduct an area-wide earthquake training exercise, stretching over five counties and lasting for four days, that involved hundreds of fire service personnel. The training began with single company responses and gradually escalated to involve the five regional urban search and rescue (USAR) strike teams operating at numerous structural collapse scenarios. The next step in the exercise was to turn over operations to the federal USAR team, who then conducted training operations for the final two days of the exercise. As you can imagine, the planning for this type of training is the culmination of years of preparation and integration of rescue teams and assets.

Rescue Equipment

You've decided which technical rescue services will meet your community's needs; selected a group of highly motivated, dedicated, and skilled firefighters to attend technician level training; and trained your support people to the awareness and operations level. Now what equipment do you need to purchase to place your team in service? Where are you going to store this equipment and how will you get it to the scene of a technical rescue incident?

Once again we begin with a community survey to determine what assets may be available. Are there lumber yards or large hardware stores in town that can lend you space? Do the local utility companies have access to vacuum trucks? Is there a local utility work contractor that has shoring equipment? Get out and talk with these people and forge agreements for them to respond to assist you with rescue incidents. Have your governing body draw up a mutual aid agreement with these outside resources, even if you have to include some sort of reimbursement guarantee.

It has taken our department the better part of a decade to acquire the rescue and support equipment and apparatus that we feel is adequate to initiate almost any technical rescue operation. Through community assessments, we began by expending funds for equipment to deal with those incidents with the highest probability of occurrence in our area. We then budgeted a small amount each year to purchase, acquire, or replace our needed rescue equipment, personal protective equipment (PPE), and apparatus. If your funding is limited, make a long-range plan that prioritizes equipment to match your community's level of risk.

Plan accordingly and then continually reevaluate your plan and progress, adjusting it as needed. Now that we have obtained the rescue assets in our initial plan, we are in the process of reevaluating our operations. We want to begin consolidating equipment and apparatus to better reflect our operations and the changes presented by regional assets that were not available when we began our plan.

This process may take several years, but the benefit will be a better prepared department that can handle any technical rescue situation. Remember, it's a marathon, not a sprint.

Apparatus

Now that we have completed our community survey, polled neighboring departments, obtained the necessary training, and purchased rescue equipment and PPE, we turn our attention to storing it and getting it to the scene (fig. 8–5). Consider these options:

- Dedicated rescue apparatus
- Rescue/pumper apparatus
- Equipment trailers pulled by service vehicles
- Portable moving containers or pods and transport vehicle

Fig. 8–5. Additional technical rescue equipment requires additional storage space and a means to transport it to the scene. (Photo by Steven Heidbreder)

Once again you should perform a survey of your department's apparatus. What space is available for a large amount of rescue equipment? Does this equipment need to be on a front-line apparatus that responds to alarms daily or would it make more sense to have specialized apparatus or trailers that only respond to technical rescue incidents? Following is a basic list of rescue and support equipment you may need:

- Water rescue equipment
    - Swift water rescue personal protection equipment (PPE), flotation devices, throw ropes, hose inflator, and other hardware
    - Ice rescue suits and rope bags
- Rope rescue equipment
    - Ropes of various lengths
    - Hardware, anchor straps, edge protection
    - Harnesses, helmets, gloves
    - Stokes and stretcher riggings
- Confined space rescue
    - Supplied air respirators and hoses
    - Air manifold to supply respirators
    - Tripod
    - Air monitoring and ventilation fan
    - Lockout/tag out equipment
- Trench collapse rescue
    - Trench panels
    - Lumber of various sizes and lengths, 2×4s, 4×4s, and 6×6s
    - Plywood for ground pads
    - Trash pump/dewatering device
    - Shovels and buckets
    - Carpentry equipment such as hammers, saws, nails, saw horses, framing squares

This is a brief list of items needed for various rescue disciplines to illustrate the large amount of equipment that will be stored and transported to a rescue incident.

My department has approached this dilemma in various ways over the past several years. We began with a dedicated rescue company and apparatus that carried almost all equipment we needed for a technical rescue incident. As the scope of our technical rescue services expanded we quickly realized we needed additional space for equipment such as trench panels, lumber, and shoring materials. We were fortunate to acquire a beverage delivery truck, which we converted into a support unit to carry all of our shoring equipment. This unit would respond as a second apparatus with the rescue company when dispatched to a collapse incident.

Recently, due to economic constraints, the rescue company was disbanded and personnel were reassigned to increase staffing on three engine companies. Since we no longer had a dedicated rescue company, we had to take a different approach to delivery of rescue services. After much discussion and research we decided to proceed with a rescue/pumper apparatus with support apparatus carrying the bulk of our equipment. Specifications were developed for an apparatus that could function as an engine company while still carrying enough equipment to begin support of a technical rescue. The current rescue apparatus was retrofitted in house to be a support apparatus and carries a majority of our technical rescue equipment.

The new rescue engine would serve as the rescue company for the district and is equipped for the operations that occur most frequently:

- Heavy vehicle extrication
- Rope rescue
- Water and ice rescue (fig. 8–6)
- Two team confined space rescue
- Truck company operations at fire scenes

Fig. 8–6. Firefighters during practical evolution of ice rescue training. (Photo by Steven Heidbreder)

The retrofitted rescue apparatus became our urban search and rescue (USAR) support vehicle, carrying the associated equipment:

- Carpentry tools for shoring and trench
- Initial stabilization equipment for trench collapse (panels and struts)
- Torches, jackhammers, and concrete saws for breaching and breaking
- Air cascade and compressor
- Extra confined space equipment (tripod, rescue stretcher, supplied air respirators)
- Aviation rescue equipment for helicopter operations
- Extra rope rescue equipment
- Light vehicle extrication
- Mass casualty incident support equipment

The beverage delivery truck expanded its role and carries shoring and decontamination equipment:

- All lumber and shoring material
- Bulk foam supply
- Emergency decontamination equipment
- Decon support equipment for the county hazmat team

Our department administration allowed our dive team to convert an ambulance being rotated out of service for use as a dive vehicle. This allows a central place for all dive equipment to be stored, ready for immediate use. In addition, we have a climate controlled space for the divers to change and rehab during extended operations.

The rescue engine will respond with the appropriate support apparatus or staff depending on the type of incident. All rescue apparatus are centrally located in one station in the district. The company officer of this station serves as the shift training officer and is in charge of rescue operations for his or her shift.

In addition to the rescue assets described above, we have set up two engine companies as engine/rescues with specialization for their particular areas. Both of these apparatus are standard engines with extra space for rescue equipment. One engine/rescue is remotely located and has an increased potential for serious vehicle accidents. This company has a beefed up complement of extrication equipment and can also serve to back up the rescue engine if needed.

The second engine/rescue is located near a large state park and river. This company is outfitted with extra water rescue, rope rescue, and wilderness rescue equipment. This company can begin operations and then be supported by the rescue engine company upon their arrival.

Presented with our rescue challenges and budgetary constraints, this approach has worked best for our organization. This is of course just one way to approach the situation and an example of a custom planned technical rescue response. Some of the circumstances are unique to our department while others are quite common in the suburban fire service. Through research and training your department will be equally well equipped to tailor a program that can best serve your specific needs.

Looking ahead to the future we are continuously evaluating our services, equipment, and apparatus. We are in the beginning stages of evaluating the possibility of converting all of our support apparatus to a portable storage container or pod system to reduce vehicle inventory and to free up space in our stations. We hope to realize annual financial savings from reduced fuel, insurance, and maintenance costs by going from three motorized support apparatus to one truck with various rescue pods. Again, this is an excellent example of the fluidity needed to achieve any technical rescue program. There is a constant need to adapt the plan to changing circumstances, be they financial, geographical, or related to new or changing needs of the community.

Summarizing the Solution

The decision to provide technical rescue services to your community is the first small step in a never-ending journey of training, evaluating, and revising your rescue operations. Develop your long-range plan from the various community and department surveys that you have conducted. Short of receiving large grants, equipment purchases will have to be spread out over many years. Start the budget process now and earmark as much money per year as your budget will allow. Hold raffles, bake sales, rodeos—whatever it takes to raise the needed funds. Begin by focusing on equipment to deal with rescue incidents with the highest probability of occurrence in your area. Expand each year until your cache of rescue equipment meets your community's rescue needs. It took our department many years to acquire all of the equipment we identified in our initial plan.

While budgeting and justifying funding requests to your governing body may seem to take every minute of your time, training should be your main focus once you have started your technical rescue journey. Since you have identified your rescue needs, you can concentrate on obtaining initial training in those areas. Use the road map provided in chapter 2 of *NFPA 1670* to help you with establishing your team and training requirements.

Remember, the initial training is only half the battle. You need to keep your team's skills and knowledge sharp. This is no easy task! Our department attempts to provide monthly technical rescue training to all team members, covering the various rescue disciplines. Logistics for some of the more specialized training can present some unique challenges. Where do you find concrete for breaching and breaking operations? If you don't have a training facility, where can your people hone their rope rescue skills? Where do you train for swift water or ice rescue emergencies? A little ingenuity and out-of-the-box thinking will help you identify locations for training events. Get your people involved in developing the training.

One more important topic not directly addressed previously in this chapter is leadership. To begin with, there must be buy-in from the highest levels of your department. If chiefs and the governing body are not going to support you and your efforts, you will be in for a long and arduous journey. Once you have gotten buy-in from your superiors, someone needs to take the responsibility of leading this group of rescue technicians. It doesn't have to be one person. Have a group of people, a committee, or whatever will work for your department assigned to keep the team moving forward.

I have found through my 20 plus years of service that those who step forward to be members of the technical rescue team are some of the most dedicated, motivated, ingenious, and gung-ho members of the department. Harness their potential, point it in the right direction, and hold on for the ride! To quote the 19th century labor politician Alexandre Auguste Ledru-Rollin:

"There go my people. I must find out where they are going so I can lead them."

constructed primarily of engineered lumber. According to a recent Underwriter's Laboratory test, these lightweight structural elements have an average collapse time of 6 minutes. Compare this to older construction techniques and solid lumber structural members used in buildings in many urban areas. The legacy construction of these buildings withstand approximately 14 additional minutes, permitting the search team to conduct interior searches that may be impossible in lightweight engineered structures. Therefore, with fewer resources and less time until collapse, it is essential that suburban fire departments train on the decision making process to quickly determine if the first arriving company should commit to fire attack or search.

Rescue Priority

Every crew has to make the same critical decisions, but few suburban crews have the resources to complete fireground tasks simultaneously. The use of SOGs and a first arriving matrix can help the company officer make decisions about task priorities more quickly and safely. We know there are no absolutes in the fire service; every fire is different with different critical factors presenting as we arrive. A simple way to put it into perspective for short staffed crews is, "If you can't accomplish fireground tasks (i.e., fire attack and search) simultaneously, then you will almost always have to choose to attack the fire first." Simply stated, if you put water on the fire the problem goes away. By choosing fire attack first, the main goal is placing water on the hostile fire to extinguish the hazard and save lives in this process.

Rescue or Line Deployment Matrix

Rescue	Line Deployment
Exact location of victims is known.	Unknown location of victims.
Exact, minimum (1 or 2) number of victims is known.	Unknown number of victims.
Extensive fire conditions.	Large number of trapped victims.
	Unknown location of fire, or it prevents access or egress.

Choose Rescue or Primary Line deployment when not enough firefighters are available to execute each simultaneously and the above conditions exist.

Fig. 9–1. Rescue versus line deployment matrix. (Illustration by Erin McGruder)

The rescue versus attack decision-making matrix for first arriving sheds light on this process (fig. 9–1).

Now let's put a couple of these decisions into perspective with examples.

1. *It is 11:00 a.m. The first arriving engine in a small suburban or rural department arrives with a four-person crew to a two-story, balloon frame house. They see at least one room post flashover on the first floor A/D corner. They are met by a hysterical female screaming that her baby is upstairs. She points to the window on the A/B corner of the second floor. The next due truck company is a mutual aid department that is en route, but 10 minutes out.*

   When the crew arrives, they are aware of the exact locations of the victim and that there are fewer than two victims in the structure. Therefore, according to the first arriving decision matrix, the correct decision would be to throw a ladder to the second floor window to the known location of the victim. The crew would then begin the vent, enter, and search (VES).

2. *It is 5:30 p.m. The first arriving engine in a small suburban or rural department arrives with a four-person crew to a four-story garden apartment. The conditions presented to the initial arriving officer are one apartment fully involved, with heavy smoke on the remainder of the first floor. There are between six and eight trapped people on the second floor landing above the fire.*

   There are more than two victims; therefore, the crew of four could not effectively rescue all victims. Ultimately they would begin fire attack to extinguish the fire, thus saving the most lives.

Using this rescue matrix is just another way to help the suburban officer make critical decisions with fewer resources.

Vent, Enter, and Search (VES)

A comprehensive effort to vent a specific window before entering that window to begin a search for trapped victims (fig. 9–2) is commonly known as a VES. This tactic is not suitable for every situation, but it is the fastest and most efficient means of reaching the known location of victims.

Fig. 9–2. Crews vent a specific window before entering to search. (Photo by 5280 Fire)

VES was first developed and documented as a tactic by a large, East coast urban fire department and has been very effective in saving countless lives in the past 20–30 years. Many urban fire departments use this routinely, allowing their dedicated truck companies to search on the upper floors while their engine crews are attacking the fire simultaneously on the first floor. Because suburban or rural fire departments usually do not have resources to attack the fire and perform a VES, they shy away from using this tactic. However, if used properly, a VES accomplished by one or two firefighters is a great tool for short-staffed fire departments.

Urban departments can use VES even without having a known location of a victim because they have an engine company attacking the fire at the same time the truck crew goes above the seat of the fire. If a suburban department does this and don't have a known victim location, they will put themselves at great risk and allow the fire to take control of the entire structure during the search. Therefore, if VES is to be used, the crew must know the exact location and number of victims present in order for it to be worth the risk of allowing the fire to go unchecked by an attack crew.

Vent, enter, and search is a very controversial subject in the fire service, so we will not began to address all the different methods that can be used to accomplish it. However, the following is a basic rundown of how to conduct a VES and considerations to include in your own department's standard operating guidelines (SOGs) or policies.

How is VES performed? VES can be accomplished from aerial ladders, ground ladders, a porch roof, fire escape, or windows that are above the first floor. The VES procedure is as follows:

1. Once you have the known location of the victim, access the window and clear the glass from the entire window.
2. Sweep the interior below the window for victims.
3. Sound the floor with your tool.
4. Make entry.

Entry for VES is very controversial, but I will explain two methods for your department's consideration when drafting guidelines or policies.

Head first entry

I personally prefer this method. Upon sounding the floor, enter the room head first (fig. 9–3). This puts you in the work position below the smoke and at the coolest point in the room. Once fully into the room, put your head to the ground and scan for the door, beds, and other important landmarks. Proceed to the door immediately. Do a quick scan and sweep within an arm's reach outside of the door, then *close the door*. Scan the room to reorient yourself to your entry window. Complete the search of the entire room back to the window and exit through your entry/egress point.

Fig. 9–3. VES can utilize a head first or leg first entry. (Photo by First Alarm Production Fire Ground Photography)

Leg first entry

After sounding the floor, enter the room by placing one leg over the windowsill, sliding back quickly to immediately drop your midsection close to the window. At this point, you should have one leg inside the structure and one leg outside the structure. The remainder of the method follows the same procedures as the head-first entry.

- **Close the door.** This is the most important step of all VES procedures. Failure to close the door will result in a very dangerous situation where the firefighter has opened a vent flow path. This puts the firefighter between the fire and where the fire is going. Therefore, it is essential to close the door immediately, buying the searcher more time to search the room.
- **The search.** Although we have covered these points above, once the door has been closed, the searcher will reorient to the entry/egress point and then begin searching back toward the window with special emphasis on places such as beds, cribs, and couches. Once the room has been searched, the searcher exits the room via the same window entry point.

Under no circumstance should you combine the VES tactic with a positive pressure ventilation (PPV) operation. Mixing PPV and VES is deadly. The fans used for PPV will push the fire and all products of combustion right out the vent opening that was created for the VES operation.

In buildings of newer, lightweight engineered construction, the weakest area of the floor beams is the front door area, which is also the heaviest loaded area of the floor beam. This is why the majority of firefighter line-of-duty deaths (LODD) happen when the firefighter makes entry through the front door and falls through the floor. This makes VES an alternative entry point, keeping personnel away from the overloaded front door area.

VES should be used with caution; however, it is an efficient life saving tactic that can be used with limited staffing and should not be discarded by suburban departments. Suburban departments need to train and be prepared to use it more now than ever before.

Searching by Occupancy

Most textbooks and classes focus on searching strategies such as primary and secondary searches. They also primarily teach right hand and left hand searches. Right hand and left hand searches are great and work well for residential occupancies; however, suburban fire departments have to learn how to effectively search other types of buildings with the resources they have available.

One way of being more effective is searching by occupancies. This technique involves preplanning for the occupancies in your jurisdiction by matching up search strategies with the key attributes of each occupancy type. In order to search by occupancy, you need to understand four basic search methods: standard, large area, aisle, and oriented.

- **Standard searches.** This is the traditional right or left hand wall search. The traditional way of accomplishing this search is for one firefighter to crawl along the wall, using the wall for the orientation point (fig. 9–4). The second firefighter maintains contact with the leader (often by holding the leader's boot) and sweeps out away from the wall to search the inner area of the room.

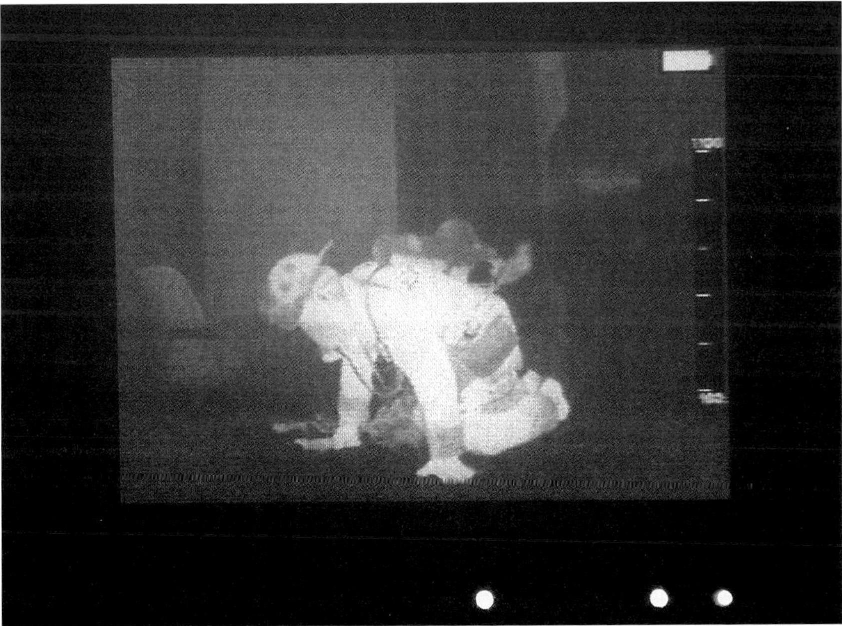

Fig. 9–4. Standard searches utilize a wall for an orientation point. (Photo by 5280 Fire)

- **Large area search.** In large area search, a rope is primarily used as a guide. This type of search requires a minimum of three firefighters. Firefighter one must secure the rope to the exterior of the building. This firefighter proceeds into the building or large area. The searcher advances the line into the building and ties a loop at predetermined point per departmental guidelines. Once the loop is tied, the two following firefighters will proceed to the loop and then hook their personal taglines in. Then the two searching firefighters should proceed back to the starting point. Each firefighter spans out the entire length of the personal tagline, searching perpendicular to the main guide rope. These searches are used for large open spaces such as school gymnasiums, factories, and storerooms.
- **Aisle search.** This search is used in commercial occupancies that have rows of aisles. In order for this search to work the company must have preplanned the occupancy and have an idea of the layout. This search is accomplished by a minimum of three firefighters. One firefighter will be the lead point firefighter with a guide rope or preferably a charged hand line. All three proceed to the aisles, where the lead point is positioned at the aisle end and one search firefighter moves down along one aisle and the other search firefighter is in the adjacent aisle. This way, two aisles are being searched simultaneously. Each search firefighter sweeps down one side for the entire length of the aisle before crossing to search back to the point firefighter. This is the quickest method of searching large commercial occupancies that have common aisles.

  Note: When entering any large commercial occupancy, firefighters should not proceed more than 150 ft deep into the structure. This will ensure they have adequate air supply to exit in an emergency situation.

- **Oriented search.** An oriented search can be used in many applications, such as in dormitories or small apartments with a center hallway. This search requires a minimum of three firefighters. The oriented firefighter proceeds down the center hallway and focuses on the areas to be searched, the fire conditions, and the safety of the crew. Once areas to search have been located, the oriented firefighter will lead a searching firefighter into the room or apartment to conduct a right or left hand search. The oriented person will direct the second searching firefighter to search the apartment directly across the hallway at the same time. Using this strategy, two areas are searched simultaneously. The oriented search can be modified in many ways to search churches, schools, nursing homes, and other occupancies that have a common layout. Oriented search is the preferred and safest method of searching because one person is oriented and focused on the safety of the entire crew while the other two physically conduct the search.

Once suburban fire departments know the four basic methods of searching, they should preplan their jurisdiction and match each occupancy type with the appropriate search method.

Hoseline Searches

Putting the first hoseline in place to extinguish the fire is usually the most successful tactic used with limited staffing. Therefore many engine crews will find themselves searching with a hoseline; especially when a dedicated truck crew is not available. Some may think this is no big deal; however, most crews are taught how to do right and left hand searches using the walls as a way maintain their orientation. When crews using hoselines began to search, they often revert to how they were trained and began conduct the wall search with the hoseline. By doing this, the hoseline can become tangled or wrapped around furniture. When the nozzle returns to the door, the hose is hung up. Crews need to train on searching with hoseline so they are successful when this tactic is needed.

The best method of conducting searches with hoselines is to modify the oriented search. Have the nozzle firefighter act as oriented person and stay at the door of the room while the two remaining firefighters conduct a search of the room. The nozzle person should remain oriented to the fire conditions. By doing this they have a hoseline that can be used for extinguishment and a way out of the structure.

Another strategy to search with a hoseline is to use it as a guide on which firefighters connect personal taglines and then span perpendicular to search a large open space.

If a crew is using a charged hoseline while searching, they need to realize that it will slow them down. Therefore, they should target areas of high probability for victims such as behind doors, hallways, beds, and around windows. Crews should check these areas as they put the hoseline between the victim egress and the fire. Although we know using the hoseline will slow down the crew significantly, it also has a huge advantage to a limited staffed fire department because the hoseline can also be used to extinguish the fire or keep the fire in check while the primary search is being completed.

The traditional truck versus engine work has become blurred as many suburban departments are expected to conduct the same number of fireground tasks with fewer resources and staffing. As crews have to adapt, they should embrace the concepts of standard operating guidelines and responses without the aid of trucks by using thermal imaging cameras and searching with hoselines (fig. 9–5).

Fig. 9–5. Thermal imaging cameras can be utilized to improve search efficiency during conditions of diminished visibility. (Photo by 5280 Fire)

Although these ideas don't match up with procedures outlined in many urban-based strategy and tactics textbooks, it is imperative that limited staffed fire departments conduct their day-to-day business with safety as the top priority. These ideas are just a few that can make the suburban fire department more safe and efficient.

Chapter 10

Rapid Intervention Teams, Mayday, and Fireground Survival for the Suburban Firefighter

This chapter was contributed by Lieutenant Scott Hulsey of the Metro West Fire Protection District, St. Louis County, Missouri, a member of the USAR Missouri Task Force-1 and lead instructor for rapid intervention at the St. Louis County Fire Academy.

Necessity and Attitude

Why do we need a rapid intervention team (RIT)? The attitude of many firefighters and fire department administrators throughout the United States is, "We don't need a RIT team; it's never going to happen to us."

Why do we have this attitude? Do we think our teams are invincible? I think the answers are easy. As firefighters, we are trained to help citizens during any emergency handed to us, including medicals, technical rescue, and hazardous materials. We don't think of ourselves as the ones getting into trouble. This is where we need to change attitudes.

To change this attitude, the change needs to start with *you*! We have to learn from those who have gone before us, those who have unfortunately given the ultimate sacrifice, so it does not happen to us. The National Institute for Occupational Safety and Health (NIOSH) reports are there for us to read and learn from. It is easy for firefighters to ask, "Why didn't they do this?" or "What were they thinking?" When we read these reports of our brothers and sisters who have lost their lives or stories about the close calls, we need to put ourselves in their shoes and ask, "What would I do in their situation?"

Definition

What is a rapid intervention team? We can start to answer this by understanding the definition of *rapid* in this context, which is the ability to go to work at a moment's notice. Rescuing a firefighter will be anything but rapid and to be successful will require everyone involved to provide hard work, quick thinking, good fireground knowledge, and some luck. But that pretty much entails everything we do on a daily basis with our jobs. This also includes dealing with a critically ill patient on the ambulance, performing vehicle extrication, fire suppression, and even our daily tasks.

Considering the definition of rapid, then what is the definition of a *rapid intervention team*? After reading several articles and books, this is what I came up with: *Rapid intervention is a team of fully equipped firefighters assembled at an emergency scene with the sole purpose to be ready and available to assist other firefighters if needed.*

Depending on where in the country you work, there are many names for the rapid intervention team. These include rapid intervention crew (RIC), firefighter assist and search team (FAST), and rescue assist team (RAT). The term RIT is prevalent here where we work in St. Louis County.

There are also different ways the RIT team can be assigned at a working house fire. The RIT team can be assigned by the incident commander, designated by arrival on scene, or designated as additional units and dispatched to confirmed working structure fires. If information is relayed that the fire is a false alarm, the unit can quickly return in service.

There are advantages to dispatching the RIT team separately from the first alarm response. First off, this truck knows once it is dispatched that they will be the RIT team. They will be able to mentally prepare and have the proper attitude to perform the RIT functions, if necessary. No one likes to be the RIT team; it is a thankless job when not utilized. As the RIT team, you are standing in the front yard, listening to radio traffic, watching crews enter the building and do the job we all love to do—fight fire. Then, once the fire is over and the RIT is no longer needed, the crew will probably be used to roll hose and help get the units back in service, listening to the other crews talk about the fire and be razzed that they were the unlucky ones to watch the fire go out. We all know this; it is the nature of the fire service to back each other up.

Then again, when you are part of the RIT team and are in the front yard listening to the radio traffic, and you suddenly hear the words that make our heart rates jump . . . "MAYDAY, MAYDAY, MAYDAY." Now it is time to do the job we all train for and hope is never needed: to save firefighters in need (fig. 10–1).

Fig. 10–1. RIT is a job we train for but hope we never need. (Illustration by Fire Medic Art)

Firefighter Safety Considerations

Why is there an average of 100 firefighter line-of-duty deaths (LODD) per year (fig. 10–2)? Why have the numbers remained consistent and not reduced dramatically? Let's try to answer these questions.

Fig. 10–2. In memory of Ryan Hummert, who answered the final alarm on July 21, 2008. (Photo by David J. Dubowski)

Start by looking at how houses are being built now; they are essentially disposable. If there is a fire, it is usually cheaper to tear down the entire structure and build again. The first thing that comes to mind is our worst nightmare: lightweight truss construction and lightweight wood I-beam or silent floor joists. Everyone wants to save money and has adapted the foam wrap house insulation and energy-efficient doors and windows. A fire inside one of these sealed houses, with essentially no air coming in from the outside, will be waiting to attack with one breath of fresh air. Where does this fresh air come from? You guessed it; we bring it with us by making entry, breaking a window, or forcing a door. The fire gets a breath of fresh air, possibly leading to a catastrophic event.

Firefighter safety is also affected by our equipment and personal gear. Consider the cost of a full set of personal protective equipment (PPE) and whether it is really keeping us safe or is a hindrance in our safety. It can easily run from $2,000 to $4,000 dollars, but think about our PPE compared to what we wore 10, 15, and even 20 years ago. Take a look at your gear and think about it. The vapor barrier liners of our bunker coats and pants are supposed whisk the moisture off our bodies to keep us cooler and dry to avoid steam burns. The gloves we wear are more durable with better dexterity and are more insulated to protect us from the heat. Our boots are lined with leather to protect us from getting wet and from heat.

The composite helmets protect us from falling debris and are designed to prevent injury and concussions. The Nomex hoods are thicker and longer to go to the base of our necks instead of just being satisfied that the hoods go down to our coats.

Now let's look at our self-contained breathing apparatus (SCBA) and what it is doing for us now. When I started as a full-time firefighter in 1993, our SCBAs were 2,216 psi with a mix of composite bottles and steel bottles. Now, all bottles are composite and are 4,500 psi. While working, our heart rates become elevated and our breathing increases, which make it nearly impossible to get 45 minutes out of a bottle. The amount of air and time is reduced roughly in half. Do you work until the low air alarm and vibrator alert starts to go off? We are all guilty of this unsafe practice! I am just as guilty as the next person.

As time and technology changes, we as firefighters need to change our thinking and become aware that we need to be outside of the immediately dangerous to life and health (IDLH) atmosphere before the low air alarm starts to go off. All firefighters have the mentality that we are here to do a job and will not quit until the job is finished. The purpose of the low air alarm is a personal insurance policy. This low air alarm tells us we have hit our reserve air. The reserve air is for our escape. Just as in confined space, where we wear the small personal escape bottle in case of a failure of our primary air supply, that is how we need to start thinking about our low air alarm. When the low air alarm starts to ring, we should already be out of the building.

Preparation

In what type of structure are firefighters most likely to get into trouble: a residential or commercial structure? This topic can be debated for hours and days. If you read the NIOSH reports and trade magazines, most are incidents within residential structures. Fire departments usually allow their companies to perform walk-throughs or annual fire safety inspections for commercial structures. This allows crews to identify the hazards, fire department connections, and occupancy patterns. In contrast, residential structures are a much more complex monster for us to understand. We do not have the ability to do safety inspections or identify the hazards in peoples' homes.

There is no way for us to arrive at a residential fire and know the layout of the house and any hazards that may exist. We can guess a general idea of layout and bedroom locations by considering the locations of windows, but that is an educated guess at best. Homes are remodeled on a regular basis, often without the proper permits, which can cause utilities hazards. Similarly, hoarders can unknowingly create significant hazards and challenges as they fill their homes with clutter.

Although we cannot control or know lies in wait for us inside each structure, we can control our safety just outside the point of entry. There isn't a safe reason to enter a working fire structure without having a designated RIT. If we think about it, RIT is a fairly new concept to the fire service. I can remember a fire in the early 1990s in which I was part of a RIT, waiting at the command post with our tools ready to go. The incident commander gave us an interior attack assignment and my company officer told the IC that we were RIT. The response we received was, "I don't care if you are RIT, I told you to go help put the fire out." So what do you think we did? That's right; we went in and helped put the fire out. Look how far we have come in 15 plus years; we now have RIT established, outside, on every fire.

There is sometimes still resistance with training on RIT and on calling a Mayday. This occurs occasionally not only in our department, but in departments all over the country. The common themes are "It's not going to happen here," "We have never had a LODD," and "We don't run that many fires to need RIT or to learn to call a Mayday." Well they are right on one point: we do not run that many fires. But not running many fires can cause complacency, and complacency is a factor that leads to firefighter deaths and many "near miss" situations. In May of 2002, the St. Louis Fire Department lost two well-respected veteran firefighters in a commercial fire. That tragic day in St. Louis has made RIT what it is today: a necessary, trained, and methodically planned function on every structure fire.

Resistance and Compliance

Why is it a hard concept to commit a RIT at all working fires? For other situations it is not an option. If you have divers in the water, then you have divers on shore ready to go. If you have rescuers in a confined space, you have rescuers at the point of entry ready to go. If you have crews operating at a hazmat scene in suits, you have crews in suits ready to go.

In 1992, the National Fire Protection Agency (NFPA) initiated the development of *NFPA 1500*, the fire department occupational safety and health standard. It is a requirement that fire departments provide a RIT for the rescue and safety of crews operating. In the *NFPA 1500 Handbook*

1993 edition, author Bruce Teele stated that "firefighters are most exposed to the risk of death or injury at emergency incidents and that using RITs is an effective way to reduce this risk.... Case studies have repetitively shown that when members learn of others in distress, everyone rushes to assist at once, causing confusion and a lack of coordinated effort. RIT can be called on immediately with their only mission to rescue personnel in trouble."

In 1995 the NFPA developed *NFPA 1561: Standard on Emergency Services Incident Management System*. In this document, chapter 4.1.8 states

> *"Fire Departments shall provide personnel for the rescue of individuals operating at emergency incidents if the need arises. A Rapid Intervention Team shall consist of at least two members and be available for the rescue of personnel if necessary."*

Even with these existing NFPA standards, departments across the country and here in the St. Louis area are slow to adopt—and in some circumstances even refuse to accept the need and use of—rapid intervention teams.

Suggested Operating Considerations

Throughout this book, you have read about suggested operating guidelines (SOGs). A common theme in this book and in this chapter is marrying crews together to work as one to achieve the same goal. An example of SOG implications for RIT would be to detail the fifth arriving company as RIT. This does not matter whether it is a mutual aid company. Mutual aid units should have a working knowledge of the district SOGs, as there may be times that the incident commander may have to assign RIT to a mutual aid company.

As mentioned earlier in this chapter, it is common to have our suburban fire apparatus run with only three firefighters to a working structure fire. This is where marrying companies together comes into play, making two companies into one. You could even consider the concept of having a rotating or *on-deck* crew that is ready to deploy (fig. 10–3). This concept can be described as having your RIT team ready at the command post and having a crew that is ready to move up to the RIT team if your initial crew is utilized.

I know some may be thinking, "How can we have that happen? It's a waste of our resources to have too many crews on the outside waiting to do something." Is it really a waste of resources to have crews outside making the fireground safer? Think of RIT as an insurance policy for us; RIT is there if needed.

Fig. 10–3. The RIT stands by on the outside of the structure as the primary interior attack is underway. (Photo by Box Alarm Production Fire Ground Photography)

Once on scene and designated as RIT, all members of the company, in full PPE and with SCBA, will report to command for a staging assignment. The company officer may direct his crew to get tools and report to the command post, but he bears the responsibility of crew integrity and to conduct a 360° evaluation of the structure.

This evaluation will answer several questions. Where did the initial crews enter the building? Where are the crews now operating? Is this a single story residential home, or from the back is it an atrium ranch or two-plus stories? The officer will then come back to his crew and relay the information to his crew.

Equipment

The equipment that RIT should bring to the staging area will vary depending on the building, location, and type of operations that are ongoing. Consider the minimum equipment that should be available (fig. 10–4):

- Thermal imager
- RIT bag with search rope
- Flathead axe
- Halligan
- Pike pole

Some may argue that you need to take more equipment with you to RIT staging. Sure, you might need more equipment than stated above, depending on what type of building you are dealing with. There may be instances where you will need a chain saw, partner saw, or other specialized equipment. You might take all of the specialized equipment, including air bags and pry bars. But realize that getting this equipment off your trucks takes multiple trips back and forth and can waste valuable time away from getting your crew to the staging area. Why not have the on-deck or backup crew get a few of these specialized pieces of equipment? This way, your primary team can be ready for action as soon as they are needed. Hauling much more than the recommended equipment might exhaust the RIT crew or require an additional truck. The equipment listed above is the basic and minimum to be considered for possible RIT operations.

The primary function of RIT is to make the fireground as safe as possible for the operating crews. This can include placing ladders as a means of entrance and egress for interior crews, shutting off the gas to the house, and cutting bars off doors and/or windows for easy access as needed.

All RIT members need to have a portable radio on their person to monitor radio traffic and keep track of the locations of interior crews (fig. 10–5). In fact, all members that are working on the fireground need to have portable radios. Every piece of Metro West apparatus carries a portable radio for each riding position, and everyone is expected to have a radio.

Fig. 10–4. RIT tools in the staging area. (Photo by 5280 Fire)

Fig. 10–5. All RIT members should have a portable radio on their person. (Photo by Erin McGruder)

The Mayday

If there is a Mayday called, command will acknowledge the Mayday, obtain LUNAR information (location, unit, name, assignment, and resources), and then direct the primary RIT team to enter the building. Other than RIT, what do you think is the most important function on the fireground when a Mayday is called? What function needs to continue to ensure that conditions do not worsen inside the structure? If you think it is fire suppression, you are right. The other pre-Mayday functions on the fireground, such as ventilation and search for victims, are all very important. But fire suppression needs to be continued to prevent further fire spread. This one activity plus ventilation needs to be completed when a Mayday is called in order to make the environment as survivable as possible.

A crew exiting the building with low air may be tempted to return when the Mayday is called. Is this the right thing to do? Is this a safe thing to do? Is this a reality and will this happen? The answers to the first two questions are *no*. It is not the right thing to do and not a safe practice. This practice increases the likelihood that more Maydays will be called as the low air firefighters do not have sufficient capacity to search and self-extricate.

The answer to the third question, is this going to happen, is regrettably *yes*. It will continue to happen. This is a reality and it is in our nature as firefighters to go back into the building when a Mayday is called. This is why there needs to be a RIT sector or branch set up, with a chief officer at the point of entry to monitor the crews going in and the crews coming out.

The first crew to enter the building after the Mayday will enter based on the information gained from the LUNAR report given by the firefighter(s) calling the Mayday. The first RIT should enter the building with the RIT pack, search rope, thermal imaging camera, and at least a set of hand tools or irons for lifting and breaching if needed. After the primary RIT has reached the downed firefighter(s), they should advise command of the air status of the firefighter, injuries, and any additional equipment needed to extricate the firefighter from the building. Once this information is received by command, the next RIT on deck should be placed at the point of entry to help extricate the firefighter. Once the on-deck crew has been deployed, a third team needs to be placed in the same position. The first crew might extricate the firefighter, but there will be times that it will take two, three, four, or more crews.

When a Mayday is called commanders have numerous things to consider. Some items that command must do when a Mayday is called include creating a PAR (personnel accountability report), obtaining a LUNAR from the firefighter calling the Mayday, requesting additional alarms to the scene for extra RIT personnel; all of these tasks must be performed in addition to other vital fireground activities.

Calling a Mayday

As a firefighter, when do you call a Mayday? What are the parameters for calling a Mayday? What information do we transmit to the IC? Why do firefighters want to avoid calling a Mayday? All of these questions can be traced back a department's culture and training.

How often do you train on Maydays? Mayday training should be incorporated into the division's training plan at least on a yearly basis. Calling a Mayday does not have to be because someone has fallen through a collapsed floor or roof, or is lost and/or disorientated. Calling a Mayday can be for someone who has fallen ill, run out of air, been injured, or has some other emergency. A reason for me to call a Mayday may be different from a reason for you to call a Mayday.

If you feel that you are in any sort of trouble, do not hesitate to call a Mayday; it's better to have RIT on the way inside the building than waste valuable time because your ego has gotten in the way. There are no real parameters for calling a Mayday; many departments leave it up to each individual firefighter to decide when they need to call a Mayday. At the St. Louis County Fire Academy, where I teach RIT/Mayday, I usually have a mix of recruits that have some previous career or volunteer fire experience and some who are as green as green gets to the fire service. When teaching this two-day class, I tell the recruits that there are some basic parameters you need to use when calling a Mayday:

- If you or your partner become lost and cannot find a door or window within 30 seconds
- If you or your partner become entangled in wires and cannot extricate yourself within 30 seconds
- If you or your partner fall through a floor
- If you or your partner is involved in a ceiling collapse
- If you or your partner become ill (shortness of breath, chest pain, etc.)
- If you become separated from your partner and/or the hoseline

After you or your partner decide to transmit a Mayday, remember to transmit your LUNAR information to the IC:

L – Location

U – Unit

N – Your name/nature of the Mayday

A – Amount of air you have left and what was your assignment

R – Resources needed to get you out

Here is an example of calling a Mayday:

Firefighter: "MAYDAY, MAYDAY, MAYDAY."

Command: "Unit with MAYDAY, go ahead."

Firefighter: "This is Lt. Hulsey 3312A. I am on the second floor bedroom fire attack and there has been a collapse, I have 1800 psi of air left and I cannot move my leg. I am with my partner and unknown of his air and injuries, we need RIT."

Command: "Command is clear, 3312A with partner second floor bedroom involved in a collapse; unknown air supply on firefighter, RIT will be activated."

Command: "All units operating on the scene at 123Main Street, switch to command channel 15. 3312A, RIT, and command will remain on command channel 16."

The information relayed from the firefighter to the IC contained all of the pertinent information needed to send RIT to start the search and rescue process. Was the information given by the firefighter in the order according to LUNAR (fig. 10–6)? No, it was not, and it does not have to be as long as the pertinent information is relayed to command.

Fig. 10–6. LUNAR is an acronym outlining information helpful to an activated RIT team. (Illustration by Fire Medic Art)

Do not let embarrassment, fear of being teased, not knowing the parameters of when to call a Mayday, or lack of department training keep you from calling a Mayday. All firefighters can admit to the ego factor; we are the ones running into the building when everyone else is running out. We all know we have the best job in the world. We are considered brave and when we get into a situation, we always think we can get out of it without help. If we ever become trapped, we need to check our egos at the door and realize that we are human and there may be times that we need help to get out of a situation.

The training we receive from our departments range from special operations such as technical rescue, EMS, and hazmat to fire training. Most of our fire training revolves around tactics and strategies. There is very little training that goes into RIT and calling the Mayday. There needs to be more emphasis on RIT training, involving our mutual aid companies, marrying the various companies together, and working with each other so there is no freelancing. If we work together with our mutual aid departments on a regular basis, not only for RIT operations, but in all aspects of the fireground operations, we will have better results in saving life, property, and looking out for each other.

In suburbia, some departments may only run one, two, or three companies and rely on mutual aid to assist on structure fires. We all know that training is essential and more emphasis needs to be placed on RIT and calling the Mayday. This needs to be done so if the need ever arises for a Mayday call, the crew(s) that is assigned RIT will rely on instinct and in turn the firefighter(s) that call a Mayday will also rely on instinct and have the training needed to get through the emergency.

As a fire service, we have done a good job watching out for each other. We need to do a better job now and tomorrow due to the dangers that are increasing before us. What will be the next danger that lies ahead? Will roof trusses be constructed of the same material as silent floors? Will new homes be even more airtight than they are now? We have no idea what lies ahead. We need to be ahead of the game, plan for the worst, and hope for the best. This is where training on RIT and Mayday needs to be a priority with all of us.

The fire service is one big family; we look out for one another and make sure we all go home to our families and friends. The days of having a bread and butter or routine fire are long gone. With the way buildings are being constructed now and in the future we need to be aware of possible catastrophic events happening sooner in the timeline of our on-scene operations. We can look at training pictures all day long; however, the best training we can do is to go out and get hands-on experience by making connections with the local builders. This way we can see the buildings under construction and know the components of the buildings. With that said, everyone stays safe, makes every day a learning day, and remembers the credo "Omnes cedo domus." *Everyone goes home!*

Part III

Formulating a Suburban Fire Tactics Game Plan

Developing a successful game plan requires a systematic approach. This section of *Suburban Fire Tactics* will explore what exactly makes suburban firefighting different from urban firefighting and discuss strategies and key elements for implementing suggested operating procedures. The conclusion of the strategic planning model will focus on reevaluation of the implemented procedures and assist with future planning and needs assessment.

Coordinating services in order to maximize efficiency can be a challenge. It cannot be conducted haphazardly. This section is a tool that will aid in developing a complete operational plan utilizing the strategic planning model.

Chapter 11

Tactical Variables: Staffing, Available Apparatus/Resources, and Response Area

Stating the Challenge

It doesn't take more than a quick observation to realize that suburban-based firefighting circumstances differ greatly from their urban counterparts. Anyone who has any type of involvement in either operation can provide numerous examples supporting this statement. Are there major differences? How do they affect operations? Do these differences impact service delivery? Since the obvious answer to all three questions is yes, we need to compensate for these differences and operate under different standard operating procedures (SOPs) or suggested operating methods (SOMs).

This chapter explores these suburban differences in depth and explains their impact on structural firefighting and emergency scene management. To simplify this undertaking, the examples are grouped into three main categories:

- Staffing
- Available apparatus/resources
- Response area characteristics

These are the key variables that affect decision making and truly define the need for separate fire tactics for the suburban setting.

Suburbia Defined/Explained

Before the three main categories, or variables, can be explained in detail, we must picture suburbia. We must truly understand and appreciate what is happening outside the urban confines. The suburban development is a phenomenon that has had a much greater impact on America than just the application of providing fire protection and fire safety. It is a social issue that has transformed the landscape of our country and many others across the globe. The rapid development of suburbs has also contributed to the circumstances and challenges that the suburban firefighter faces today.

Definition

It is difficult to find a static definition for the term "suburb." Suburbia has been a dynamic development since its inception in late 19th and 20th centuries. The term "suburban" is usually defined for statistical purposes as "any place in the metropolitan area outside the central city." Also known as an edge city, suburbia is made up of towns or unincorporated development areas that are in close proximity to a central or core city. Suburbs are usually residential communities and characterized by low-density development in relation to the city (fig. 11–1).

Fig. 11–1. Suburbs can be defined as residential communities with low-density development compared to a core city. (Photo by Erin McGruder)

The suburban definition is changing over time, however, due to social and economic pattern shifts. Suburbs, because they were largely residential, were once considered dependent on the urban core for employment and support services. Now considerable industrial development has transferred outside of the cities into the suburban regions and their dependency on the core city is diminishing. Today, this influx of population away from the city borders has created large metropolitan areas where the combined population of the suburbs exceeds that of the core city.

Characteristics

When describing suburban communities, a few characteristics and generalizations can be made in relation to their physical makeup:

- Lower densities
- Zoning patterns
- Subdivisions
- Shopping malls and strip malls
- Cul-de-sac road networks
- A greater percentage of low rise buildings (specifically one- or two-story administrative buildings)
- Less ethnic diversity

One of the advantages of suburban areas is that they usually designate residential and commercial occupancies through zoning laws, with the commercial buildings and activities in a location separate from housing developments. These commercial areas are commonly of low rise, large box construction, predominately shopping malls and strip malls that sit behind large parking lots and typically bear little resemblance to a downtown shopping district.

The residential areas in suburbs are lower in density than those of urban areas, commonly single-family homes that are located in subdivisions. Because the subdivisions are carved from previously rural land plots, the developments normally do not follow a city grid system for streets but rather wind around natural features and end in cul-de-sacs. The homes in each subdivision are generally similar in construction and value, creating homogenous areas of family incomes and demographics.

Suburban evolution

The suburban population in North America exploded during the post-World War II economic expansion. In the U.S., 1950 was the first year in which the suburban population was larger than that of the city. What caused this phenomenon? Scholars attribute this shift to a range of variables during this era:

- Returning veterans from World War II in search of the American dream
- The development of zoning laws in the suburbs
- Numerous innovations in transportation (the automobile)
- Availability of FHA loans
- Economic growth

The mass production of the automobile and the expansion of mass transit, such as streetcars and trolleys, allowed for workers to venture out and live farther from their places of employment. State and local governments soon responded with massive road/highway building projects that gave virtually unlimited restrictions on travel. Deteriorating living conditions and dense populations, which contributed to inner city congestion, increased the incentive for suburban expansion and urban abandonment. Soon, the suburbs became the bedroom communities for the urban commercial districts. The attraction of the suburban lifestyle evolved.

America was also experiencing a large economic growth pattern that encouraged this demographic shift. Massive investments for new infrastructures and housing were required to sustain these new communities, shifting America's purchasing power. This purchasing power was becoming stronger and now more accessible to a wider range of people.

Suburban expansion did not stop with just residential living. As more population moved into the suburban ring, so did spending trends and habits. The shift of population out of the central cities has attracted industry and commerce to the suburbs. Zoning for light industry within the suburbs began to evolve. Industrial parks, shopping centers, malls, and office buildings have taken hold and have become part of the suburban landscape. Suburbs are particularly successful today in attracting newer types of industry, often including high-technology products. This new economic activity has made the more accessible suburban towns almost as large and reminiscent of some midsized central city downtown districts.

Through the 1990s, some early tier suburban communities became extremely dense and reached urban levels to the point their growth slowed and some even lost population. In many instances, populations began to spread farther from the central core of the city, expanding the suburban region. Suburbs began to sprout up that weren't even linked to a central city. Converging bands of suburbs became more prevalent with this shift in population density, rivaling other metropolitan corridors.

Discovering the challenge

Expansion in the United States is a not a new concept. Urban centers have been popping up all over North America for the last 250 years. Why do we need to state the differences now between urban- and suburban-based societies in order to develop firefighting tactics? The root of the dilemma is truly the rapid growth trends. The real question is, were the suburban communities planned for by community developers or did they simply spontaneously grow to meet population demands? If they were planned, were prediction estimates accurate? Is the infrastructure able to support its population? Regardless, suburban fire agencies are sworn to protect this population with the resources that are afforded to them.

We must take all this demographic, social, and economic information into account to understand how it influences and affects the delivery of fire protection services. A fire service agency can only provide the resources that are fiscally attainable. A majority of infrastructure growth, as previously demonstrated, is not under the control of current fire agency governing bodies. Fire service policy makers do have a role in strategic planning and can recognize the key variables in service delivery, and they can implement policies and procedures that acknowledge existing circumstances in order to derive desirable outcomes.

Staffing

The staffing of apparatus is a major determinant in developing tactical action plans. It is plain to see that without human resource capital, functions on the fireground cannot be executed. You can buy all the big and shiny fire apparatus possible, but if you don't have the people to operate them on a fireground, they are worthless (fig. 11–2). Human resources will always be the highest priority when determining tactical strategies.

When discussing staffing, an agency should be concerned with two factors:

- How many members are needed or are available
- Experience/training levels

Fig. 11–2. Human resources are the highest priority in determining tactical strategies. (Photo by Erin McGruder)

According to the National Fire Protection Association (NFPA), the minimal staffing for each fire suppression apparatus should be four firefighters. This number allows for one apparatus operator, one officer, and two firefighters, an assignment that allows for a safe initial interior attack at a structure fire. It is also a safe number in relation to effectively performing individual fireground functions. Four is ideal and the minimal standard; however, we are not so naïve to believe all agencies are able to comply with this standard. Many suburban based agencies, my own included, are not able to comply 100% of the time on each apparatus.

I agree with this standard and believe that it is for the good of the fire service. The purpose of this standard is to establish "proper safeguards against loss of life and property" while improving the methods of fire protection and prevention. As an industry, we should strive toward compliance and improvement. However, I am also realistic and realize that it is not within the budgetary constraints of every agency to attain this minimal standard. Fire service agencies are not going to shut down companies because they fall below this mark.

Options for maintaining staffing levels

If you have participated in a fire service budget workshop, you have probably learned that 90% of the budgets of all full-time, paid fire departments are allocated for human resources, which includes salaries and benefits. Unfortunately, it's a matter of simple math and budgetary constraint. Dealing with a fire department budget is not an easy task in relation to human resources. When you purchase a fire truck, the amount spent is fixed and is usually taken out of a capital expenditure account or possibly a financing vehicle, such as a bond fund. You develop an apparatus replacement plan and attempt to forecast and budget for your next purchase. Human resource spending isn't quite so simple. An employee could possibly be a 30-year investment of salary, benefits, and possibly a pension. Now, try to forecast a flow of income that will support numerous, maybe hundreds, of loyal employees over a long span of time. The point I am trying to make is that some fire agencies just cannot afford to comply with the NFPA recommendation.

The answer, which some agencies have turned to or have always used, is relying on volunteers. Unfortunately, trends have shown a decrease in volunteerism across the country. Many volunteer agencies have reported decreases of up to 50% of their force over the last 10 to 15 years. The National Volunteer Fire Council is also reporting that there is an *aging* trend in volunteer firefighters. The trend finds that the median age of volunteers is 40 years old and climbing. This number indicates a problem with recruitment among the volunteer ranks.

There are three primary types of staffing options available for fire service organizations:

- Full-time paid
- Volunteer
- Combination

All three forms of staffing can be found scattered throughout suburban America.

Full-time paid staffing. Of the 1,148,100 firefighters who protected the United States in 2009, only 335,950 were reported as paid. This number accounted for 29% of the population census and represents a clear minority. Facts collected by the NFPA also indicate that the majority of career firefighters (73%) are located in communities that protect 25,000 or more people.

A primary benefit of a fully staffed career fire agency includes the ability to ensure a consistent and timely level of response (fig. 11–3). This system guarantees a minimal level of responders at all times. There is typically a higher level of commitment due to the dependency on occupation and not competing with other employment and aspects of everyday life. This commitment not only involves emergency response, but also training.

Fig. 11–3. A fully staffed fire agency guarantees a minimum level of responders at all times. (Photo by 5280 Fire)

There could possibly be disadvantages to the full-time paid staffing model. If an organization does not have the proper funding mechanisms or a budget to sustain adequate staffing coverage, problems might arise. It is important to state the effects of improper funding mechanisms because, often in suburbia, demand for services explodes rapidly past preexisting funding. Agencies with limited resources are often forced to provide the same delivery as larger entities. Staffing "light" or insufficiently can prove to be dangerous and place members in difficult situations with otherwise avoidable critical decisions. Unfortunately, the bottom line to staffing is the budget.

Volunteer staffing. Volunteer firefighters represent 71% of all firefighters within the United States. The NFPA counted 812,150 volunteer firefighters in 2009. Most volunteer firefighters (95%) are in departments that protect populations fewer than 25,000; more than half are located in small, rural departments that protect fewer than 2,500 people. Even though these statistics are more representative of the rural based agencies, it does fall within numerous suburban settings.

Organizations depend on volunteer human resources for various reasons. The main and obvious reason is lack of funding. A volunteer system is less expensive to operate and allows for more investment in capital equipment.

The disadvantages associated with utilizing volunteer manpower are many:

- Difficult to maintain response times
- Retention and recruitment challenges
- High turnover in comparison to career firefighters
- Difficult to maintain consistent training and services
- Difficult to consistently staff apparatus due to work schedule conflicts

Combination staffing. A number of fire service organizations throughout the United States use a combination of career and volunteer staff members. The utilization and proportions of paid and volunteer members is determined by the needs of the organization. Often, volunteer members are used to supplement staffing on evenings, weekends, or during other high-demand times. Other organizations may use volunteers to back fill overtime positions created by the vacancy of career members due to sick time or vacation. Another model requires volunteers to fill a certain number of required shifts in a specified time period. A final example is utilizing volunteer manpower to supplement staffing numbers for large or special call-in incidents. These models may include incentive pay for the volunteer staffing, who are considered *paid on-call* employees.

Combination departments are designed to keep responses times minimal or comparable to full career departments while utilizing a low-cost delivery method. Manpower is supplemented while reducing human resource constraints to budgets. The advantage is the cost savings; however, there are drawbacks:

- Management challenges
- Conflicts between career and volunteer staff members
- High turnover of volunteer members
- Recruitment efforts
- Difficulty maintaining consistent training levels between all staff members

Cross-manning of apparatus

Today's modern fire service organizations have accepted much more responsibility in relation to service delivery. The biggest challenge is providing advanced life support for the communities we protect. This undertaking has provided an opportunity to both enhance services and take advantage of combination of services (fig. 11–4). This is extremely advantageous for agencies that do not have a heavy call volume, as is the case with many suburban based agencies.

Fig. 11–4. A cross-manned station prepared to provide a fire or EMS response. (Photo by Erin McGruder)

Career firefighters, in many parts of the country, are not longer *just* firefighters. Often in many circumstances they wear more than one hat. In the world of advanced life support and prehospital care, responders are firefighters as well as emergency medical technicians or paramedics. Cross training allows for satisfaction of multiple demands with a single pool of resources but may cause difficulties in a high call volume environment where only one vehicle may be staffed at a time.

Let's analyze some examples:

1. *An engine house located in a lower call volume environment is stationed with an engine and a life support unit (ambulance) and staffed with a minimum of two, but sometimes three members. At least one member is certified as a licensed paramedic and one is an officer. During medical emergencies the staff responds with the ambulance and places the engine out of service. In the event of a fire related incident, the engine responds and the ambulance is now rendered out of service.*

Advantage: In this example, there is a maximization of resources and a greater justification for services. Two units can be staffed with a single pool of resources.

Disadvantages: This model does not work well in busy environments. If an emergency medical call occurs, now fire protection is limited and there is a gap in the protection area. The opposite situation occurs during a fire related incident.

2. *An engine house located in a moderate call volume environment is stationed with an engine and a life support unit (ambulance) and staffed with four to five members. At least one member is a paramedic and one is an officer. In this example, the ambulance can respond to emergency medical incidents and still leave the engine with minimum staffing. This staffing level also allows for full compliance with NFPA 1710 standards.*

   Having five personnel provides the option of leaving the ambulance in service and staffing the responding engine with three firefighters for structure fires. In most cases where the engine is first due, however, the best option is to place the ambulance out of service by responding with a full five person engine for an efficient initial attack. These are all possible options under the cross-manning model.

Advantage: Use of resources is maximized and two units can remain staffed simultaneously.

Disadvantage: During an emergency medical response, the personnel available for the engine becomes limited. Although the engine can respond, capabilities to deliver initial fireground functions and safety have diminished.

Take the example a bit further and add the advantage of staffing multiple paramedics on the crew. Now the engine, as well as the ambulance, has the option to be furnished and operated as an advanced life support apparatus. In this case, when the ambulance is out of service on a prior response, the engine can respond and render initial advanced care before a delayed neighboring ambulance can arrive.

This last example involves the same engine house configuration in a moderate/heavy call volume area, staffed with six members.

3. *The engine is staffed with an officer and at least one paramedic. The ambulance is always staffed with two, with at least one a paramedic. In this case, the minimal manpower on the engine is the standard set by NFPA 1710. The option is still available to utilize the ambulance personnel for an initial attack. In a first due situation, the engine crew can deploy a line while the ambulance personnel perform limited facilitating (truck) functions.*

   Although fire service based EMS can be a challenge it can also be the fire service's answer to maximizing services while simultaneously answering a major community need. Prehospital care is an opportunity for the industry to gain further exposure and expand services. But what does providing this service entail and do the benefits outweigh the risks? There are definite benefits, but each organization should complete a thorough needs analysis and organizational assessment before committing to a prehospital care service (fig. 11–5). Once again, not all agencies are equipped with the resources to deliver the same services. Planning should be the first step in the process.

There are different models in relation to providing prehospital care in the fire based emergency medical service delivery:

- Providing first/initial response with first responders, emergency medical technicians, or paramedics staffed on a nontransport vehicle (e.g., engines, utility vehicles, rescue squads, etc.) for support to private or hospital-based ambulance services

- Providing an advanced life support transport system with ambulances staffed with paramedics and emergency medical technicians

Fig. 11–5. Fire service based EMS care answers a major community need. (Photo by Michael Thiemann)

- Providing a full advanced life support system that includes a prehospital transport system and complementary staffed/equipped advanced life support suppression vehicles. (This system requires at least one licensed paramedic on the suppression apparatus, which should be equipped with all essential medical equipment.)

The model a particular organization selects should be determined by its individual needs, abilities, and resources. It might be mandated by the protected community.

The following factors should be evaluated when determining the feasibility of a prehospital-based service model:

- Staffing availability and training/certification levels
- Available physical resources such as equipment and apparatus
- Separate funding mechanisms for EMS
- Call volume
- Ability to comply with medical protocols and medical control procedures established by a base hospital
- Liability coverage

One last note on cross-manning and providing prehospital care is determining call volume. Be very cautious when mandating dual job set requirements. If an organization has a heavy call volume in both medical and fire related incidents, it may not be realistic to believe that they can deliver both skill sets with equal efficiency without rotating different positions. Often, this requires rigorous training schedules for both disciplines. Skill assessments should be routinely conducted to ensure proficiency.

Mutual/automatic aid

Most suburban fire service agencies have adopted mutual aid/automatic aid agreements with their neighboring departments because of limited organizational size and staffing resources. Mutual aid is way to create seamless borders, improve service delivery in relation to safe operating methods, and minimize incident response time. Many departments rely on agreements to provide firefighting resources on both a routine and disaster basis.

In essence, there are two forms of mutual aid agreements—automatic aid and mutual aid—and it is essential to differentiate between the two.

Automatic aid is an agreement between two or more agencies for immediate response on alarms. There is no general request for aid, it is *automatic.* Automatic aid can be used for any type of alarm and can be used to back fill for any company already committed to an incident. In many cases where a progressive system is in place, agencies will dispatch the closest unit regardless of boundary restriction.

Mutual aid is the all-encompassing term used to describe the reciprocal assistance by fire departments under a prearranged plan. However, true mutual aid (distinguished from automatic aid) requires a form of request and is not automatic.

Having a solid automatic aid agreement/relationship with neighboring agencies is essential when developing suggested operating methods (SOGs). The ideal game plan established for fireground success and safety may reach past the capabilities of an individual suburban fire agency. In fact, this is true for the majority of small and medium sized departments. In order to accomplish all essential fireground functions, depending on staffing capabilities, a first alarm assignment should have no less than five pieces of fire suppression apparatus. Simple math shows that if your agency doesn't have at least five companies available, you are deficient. SOGs require coordination and timeliness. That is why the best aid agreement for this situation calls for automatic aid, as opposed to less-timely, requested mutual aid.

The benefits of both forms of mutual aid are obvious. As stated above, automatic aid may be the only solution to satisfy the ability to safely staff a complete first alarm assignment. Effective and safe operations are the main concern. Mutual aid can also allow for filling key command and general staff functions for an individual department in times of need. Mutual aid agreements satisfy many requirements established by the Insurance Service Organization, whose public protection classification (PPC) ratings could have a substantial impact on lowering insurance premiums for both commercial and residential occupancies within the protected area. These arrangements will also help satisfy mandates established by NFPA 1710 and, for volunteer agencies, NFPA 1720.

No community can handle a disaster without outside assistance. It is not feasible, efficient, or practical for all suburban agencies to commit all of their resources toward special disaster situations. It is much more efficient to have a collective pool of resources and to share the burden than to have redundancy of resources that do not fit the size of scale for a community. Examples of these mutual aid agreements/relationships are:

- Agreements with federal urban search and rescue (US&R) teams
- Local agreements with regionally sponsored urban area security initiative (USAI) teams
- Establishment of group/regional technical rescue teams or task forces
- Establishment of a state mutual aid network to pool strike forces and task forces
- Establishment of local hazardous materials teams

Successful mutual aid has many requirements, and establishing written agreements is just the tip of the iceberg. In relation to automatic aid, it is critical that all parties are familiarized with standardized policies and SOGs. These require effective communication and should include routine coordinated training events.

Agencies should have interoperability through open dialogue at both the command and staff levels. Also, crews should have interoperating radio frequencies and established fireground communication procedures. In relation to communications, a coordinated dispatching effort is always an advantage; group dispatching for regional agencies can be quicker and more efficient than separate and individual agency dispatches for aid. Even if this isn't the case, dispatchers should be trained on how to obtain resources and initiate mutual aid requests once ordered by fire department command staff.

Experience and training

The second main component of the staffing variable is the training and experience levels of personnel (fig. 11–6). Looking at the big picture of accomplishing fire department functions/activities on the fireground, we can see the value of experienced and trained operating members. Even though the common public may view some of these functions as simple and mundane, these activities require a high degree of skill and extensive knowledge to perform safely and effectively. A fire department can boast reaching and exceeding NFPA standards; however, the quantity of staffing is worthless unless those individuals can be expected to perform with proficiency and *safety*.

Fig. 11–6. Training and experience make up the second main component of the staffing variable. (Photo by Steven Heidbreder)

What is the minimal standard that should be set by a fire service organization for personnel operating suppression apparatus? The answer to that question is dependent on the planned use of each apparatus. Many of these standards are mandated by local and state regulations. If this apparatus is expected to be a vehicle that is directly related to fire suppression, the minimal requirements for personnel should be an NFPA compliant Firefighter I and II certification. These certifications require a minimal skill proficiency and number of training hours related to job specific tasks (fireground activities). Even as essential as this standard appears, there are companies operating in suburban America that are unable to fully comply.

SOGs and procedures should be written to your agency's capabilities. If your initial apparatus is not staffed with competent, trained firefighters, then you cannot safely initiate all tactical plans. As stated previously, interior firefighting operations are dangerous and require a level of proficiency, skill, and experience. If an agency does not staff each company to this minimal ability, considerations should be built in. For example, the agency may not be able to initiate interior operations with one company. They might have to wait for a second arriving apparatus to assemble the properly trained entry team. Also, keep in mind the rule of "two in, two out." Not only is the operation dependent on interior attack crews, but also available crews in reserve who are able to back up and possibly come to the aid of these interior crews. Some less fortunate agencies, lacking sufficient trained personnel, have policies that do not allow for interior fire operations unless a confirmed life is in danger. While these situations are not ideal, they are an everyday reality in some areas.

Let's take the agency capability a step further, into advanced operations. Minimal standards are exactly that, the *minimum*. Basic principles in firefighting (and most state Firefighter I and II programs) are the basic certifications and knowledge sets that allow a new firefighter to enter a structure involved in fire. Graduating from a recruit academy does not put a giant "S" on the front of your turnout coat or give you a red superhero's cape. Experience is a major component of firefighting capabilities. If your agency lacks tenure at the senior firefighting positions or officer ranks, serious considerations should be made to tactical strategies.

This concept is illustrated with the following example. An engine company arrives second due to a well-involved structure fire. The first due engine has conducted an accurate size-up and has initiated an aggressive interior attack. There is no detailed truck company, so the second due is expected to complete the facilitating and support operations that will assist fire floor operations. How many fire floor functions can this company

Fig. 11–7. Swift water rescue technician training. (Photo by Steven Heidbreder)

Putting it together: compare and contrast

To better explain the variable of manpower, compare and contrast the capabilities of an urban engine company to that of a typically staffed three member suburban company. The following illustration uses the personnel assignments of an FDNY engine staffed with either five or six members, including one officer:

- Officer
- Nozzle
- Backup
- Door*
- Control
- Chauffer

**In the event that the company is limited to five members, door and control are combined.*

Under this staffing solution, all actions that are needed to stretch a line are divided, broken down, and assigned to an individual member. The nozzle firefighter does nothing but operating the nozzle and directing the stream. The backup position drags the hose line and assists with controlling nozzle reaction. The door firefighter pulls the hose around corners, humps hose through doors, and checks for kinks. The control firefighter pulls off the desired number of lengths ordered by the officer, assembles the line, and also assists with all kinks.

Now compare the same task assignments to a suburban engine company staffed with three members:

- Officer
- Nozzle firefighter
- Chauffer/engineer

In this scenario, it is clear that all actions are not clearly satisfied. The officer is now both the supervisor and hose puller. Who is responsible for kinks and bends? If the stretch is above or below grade, this is a major challenge and is not safe.

For the hypothetical suburban example, successfully placing this attack line entails coordination. The example urban company, excluding high-rise scenarios, can complete the function of stretching the hose line with one engine company. The limited staffed suburban operation will require multiple companies combined together to effectively place a single attack line.

Available Apparatus/Resources

The second category we use to illustrate suburban differences is available apparatus and resources available for response. Typically, fire department functions are divided into those activities that are directly related to suppressing fire (stream application) and supporting or facilitating these activities. It was, and in many cases still is, traditional for larger urban fire departments to assign these functions by apparatus. In other words, engines or pumpers equipped with hoses and water are expected to perform direct suppression on structure fires. Ladder or truck companies typically are responsible for all supporting functions not directly related to suppression. If this is the accepted firefighting principle, what happens if your organization doesn't staff truck companies?

The traditional system of engines and trucks is the most effective method of implementing a coordinated fire attack. Responding apparatus are aware of the functions for which they are responsible and the combined effort of attack and support is preestablished. Since the beginning of fire protection in the United States, early fire brigades realized that their water was worthless if they couldn't reach the seat of the fire or were impeded by structural components or circumstances. Therefore, for the sole purpose of helping get the wet stuff on the red stuff, they created *hook and ladder* companies without pumps, water, or hose. So, why is there a movement in this country to either cut back or eliminate these support apparatus? Why is it hard to find a true truck company in suburban America?

The purpose of this book, more than merely asking questions and pinpointing circumstances, is to help raise awareness and account for conditions when developing operating methods and procedures, specifically when developing suggested operating methods for the optimal coordinated fire attack. To that end, it is essential for organizations that do not operate truck companies to develop a game plan or operating method to organize its companies and divide functions.

For many suburban fire service organizations, the engine company is the workhorse of firefighting operations. It is the apparatus relied upon most for delivering fireground functions. Today's engine company takes on a whole new appearance as compared to those of years past. Because the engine company (or pumper) is now expected to perform numerous functions, its design has been changed to comply with its intended use. Specific changes include the addition of aerial devices and advanced life support capabilities. Regardless, each apparatus should be purposely designed for the needs of the area it will protect.

Options that should be considered for engine companies include the following:

- Will it be equipped with an aerial device?
- Will it be considered an advanced life support apparatus?
- Will compartment space be required for medical equipment?
- Will the apparatus need to be versatile and durable to accommodate possible high-frequency, short duration responses?
- Will the size of the apparatus be affected by the response area (street accessibility)?
- Based on the response area, what size booster tank is needed?
- How much pump capacity does the apparatus need?
- How much compartment space is needed for special equipment?

- What size hose bed is required?
- How much hose will the apparatus carry and what type of hose loads are desired for both water supply and hand lines?
- Do budgetary limits dictate apparatus constraints?

Staffing issues and budgetary constraints often dictate apparatus availability. Because of these concerns, the fire service, particularly the suburban organization, has been tasked to deliver services creatively and more efficiently. The prime example of this efficiency solution is the *quint* fire apparatus, consisting of pump, tank, hose, aerial, and ladder. However, this is not the only instance of dealing with limitations. There are two notable solutions:

- The rescue engine or pumper rescue
- The engine tender or pumper tanker

The rescue engine

Many suburban fire departments have difficulty staffing a heavy rescue unit. If they decide that truck companies are inefficient due to their lack of direct suppression capabilities, then the heavy rescue unit definitely fits within their limitation parameters. The solution to the problem is either the quint concept or designated engine companies for truck work. The solution for satisfying rescue and technical emergencies for many is the rescue engine, or also commonly known as the pumper rescue (fig. 11–8).

Fig. 11–8. Metro West Fire Protection District Rescue Engine 3344 circles a cul de sac. (Photo by Erin McGruder)

The rescue engine is primarily an engine company also detailed with rescue responsibilities. To assume these responsibilities, the apparatus must first comply with requirements set by NFPA standard 1901 for engine companies. Secondly, the apparatus must be designed with extra compartments to stow additional rescue equipment. This equipment cache will vary from agency to agency, depending on individual technical rescue needs. These needs and considerations are explained in detail in chapter 8; however, we will briefly review them now to explain this concept.

Before the apparatus is designed, its intended technical rescue disciplines must be considered. After conducting a needs analysis of the response area, it may be determined that the agency will want the apparatus to store equipment for:

- Vehicular extrication
- High angle rescue
- Trench collapse rescue
- Structural collapse rescue
- Confined space rescue
- Dive rescue
- Water rescue
- Hazardous material response

If the organization decides all of these disciplines are needed, careful planning will be required in delivery of services and development of the apparatus. This is a ton of equipment! It is highly unlikely that a single vehicle can carry all of the above in addition to fire hose, water, pump, miscellaneous firefighting equipment, and ground ladders.

In many cases, the highest utilized rescue equipment will be carried by the rescue engines. Lower utilized emergency equipment will be detailed to support vehicles without full-time staff or used only for special assignments. The majority of time, the equipment that falls in this higher frequency category is related to vehicular extrication, advanced forcible entry, or rope systems for high angle. Once again, it is highly dependent on the response area. The organization also has the option to split the responsibilities between multiple companies.

Compartment space, equipment, and overall vehicle size are not the sole considerations. This type of apparatus and the responsibility assumed dictates a staffing need. You cannot expect to assign a vast assortment of specialized equipment to a regular engine company without also providing specialty training. These technical disciplines require special skills, knowledge, and abilities. Consideration should be made to staff this company with specifically trained members who are experienced and certified in the relevant rescue disciplines.

This type of apparatus has become so versatile and effective that even large urban departments staff such vehicles. One example of this is the FDNY squad concept. An FDNY squad is designed to handle extra technical rescues within its specified borough, and also function as an engine within its first, second, and third due response areas. Their design has been often imitated by other agencies looking to develop rescue engines.

The engine tender

Depending on what region of the country you are from, another specialized fire apparatus is the *engine tender* or *pumper tanker*. This is also a dual-purpose engine that was designed to satisfy two major fireground needs with one vehicle. These two functions are water supply and the obvious, direct fire suppression as an engine company. Engine tenders are differentiated from regular engines by meeting two main requirements:

- Water tank capacity over 1,000 gallons
- A large diameter discharge

This dual-purpose apparatus creates the ability to combat structure fires in areas where water mains are either nonexistent or water volume is insufficient. Instead of staffing a separate water tender (tanker), now the water, staff, and suppression equipment are all available in one unit. This extra water allows for greater tactical options in a water restricted response area. If this apparatus arrives first due, it has the ability to make an aggressive and quick interior attack while second and third arriving companies establish the fixed or permanent water supply. This water supply may be in the form of extra tank water or an established *drop tank* (temporary water basin carried by fire service apparatus). It also gives the option of having another tender to shuttle water from a permanent fixed water supply to the fireground.

Obviously, an engine apparatus designed with a capacity of 1,000 or more gallons will be larger and less maneuverable than a standard engine. Compartment space options could be considerably less and the hose bed could also be affected. There are options as to the shape of the tank; however, the height of the vehicle is almost always a factor with an engine tender. Because of the height, hose loads could be located higher than desirable and require steps or straps for access. A final consideration is overall vehicle weight and versatility. The apparatus will be much heavier than a standard engine and may require a second rear axle. Remember, water weighs approximately 8.33 pounds per gallon. Water, hose, tools, and equipment can quickly overload a truck past the specified allowable axle weights. These factors diminish the versatility of the vehicle and restrict turning radiuses and driving clearances as compared to smaller engines.

Compared to the rescue engine, the engine tender will require less specialized additional equipment. The largest equipment consideration is the drop tank (also known as a porta-tank or porta-pond). Basically this is a foldable swimming pool that is deployed at the fire scene to hold thousands of gallons of water (fig. 11–9). The engines draft this water to supply water for firefighting operations. Tenders then supply the drop tank with shuttled water from the nearest permanent water source. Other specialized equipment relates to water supply needs, such as drafting devices and water appliances.

Fig. 11–9. Metro West Fire Protection District Tanker Engine 3333 performing drop tank operations. (Photo by Steven Heidbreder)

There are disadvantages to this concept. As mentioned earlier, the versatility and maneuverability of the vehicle is much less than that of a standard engine and restricts its operations. Consider an example where the number of tenders (tankers) is regionally limited. If the engine tender is first due or committed to fixed placement at the fireground, it is no longer available for water shuttle operations. The opposite can occur, where the engine company may be needed to complete essential, time sensitive fireground functions but is committed to water shuttle/supply activities. The first example is probably more relevant due to the fact that there are usually more engines available than water tenders.

Truck companies, as stated earlier, are extremely *effective* for coordinated fire attacks. However, policy makers have determined that they might not be the most *efficient* use of an agency's limited budget resources. This has particularly been the case during economic downturns where budgets are trimmed and restricted. Does that mean that fire departments have stopped purchasing apparatus with aerial devices? The answer is a resounding *no*. Apparatus without pumps, however, have not fared well. They are usually the first units eliminated. Suburban call volume often dictates the need for increased acquisition or lower reliance on support apparatus. Because of the lower frequency of structure fires, most agencies have found that they can satisfy daily alarm needs without truck companies. Communities have found it hard to justify staffing both an engine and truck company in the same engine house.

Since the truck companies are being eliminated, does that mean that truck company functions are no longer necessary on structure fires? That answer is *absolutely not*. Many communities, including many urban city departments, have adopted the quint concept, which combines the engine company with the ladder company to form one efficient apparatus. Common specifications for a quint, or quintuple combination pumper, include the following:

- A fire pump (minimum capacity of 1,000 gpm)
- A water tank (minimum capacity of 300 gal)
- Fire hose (minimum of 800 ft of large diameter, minimum of 400 ft of either 1.5 in., 1.75 in., or 2 in. diameter)
- An aerial device (an aerial ladder or elevating platform with a permanently installed waterway)
- Ground ladders (a minimum of 85 ft of ground ladders, including two extension ladders, one roof ladder, and one attic ladder)

Additional specifications for quint fire apparatus are detailed in *NFPA 1901, Standard for Automotive Fire Apparatus*, chapter 9.

If used correctly, the quint concept and variations of the system can be efficient. But truly, it cannot be efficient unless it is effective. The effectiveness can be determined by the ability to deliver a consistent, coordinated fire attack on a regular basis. Let's analyze a generic first alarm assignment for an organization that operates with quint apparatus:

> *A working structure fire has been reported in a two-story residential. The initial assignment is detailed with five quints-engines, equally staffed. All are responding from separate engine houses.*

- How does a coordinated fire attack occur?
- Who performs facilitating (truck) functions?
- How does the quint officer make the decision to perform as an engine or truck company?

Even if this system operates one truck per region or battalion, the same questions exist. If this company is not arriving in close proximity with the first due engine, facilitating functions cannot wait for the truck company.

Variations of the quint concept exist. In this first example, a full or *total* quint concept would utilize every apparatus as an engine with an aerial device. Other variations systematically alternate engines and quint placement, sometimes known as a partial quint concept. If this is the case, the decision to operate as an engine or truck is made by both engine and quint company officers and an engine may be conducting the facilitating functions. Truly, the only difference in the apparatus is the attached aerial device. The organization with the partial quint concept, depending on key variables, can operate the same as a full quint concept. However, considerations should be written into the suggested operating methods that dictate apparatus placement; for example, efforts should be made for aerial devices to have access to the front of the fire building.

There are several key factors essential for a successfully operating quint concept:

- The quint officer must be able to recognize fireground needs in order to make decisions effectively. Specifically, choosing between truck and engine company operations.
- The organization must have a game plan that determines operating procedures to assist with this decision.
- Crews must be disciplined and stick to the game plan unless conditions dictate alteration of the plan.
- Crews must be equally proficient in both truck and engine company operations.
- Training regiments must be developed to support skills and abilities in both functions.

Availability

The final component of the resources and apparatus variable is availability. Suburban areas are typically less dense than urban areas and therefore have a greater spacing between engine houses. These larger response areas create time gaps for all arriving companies. If this is the case, the consideration should be written into procedures and the functions should be closely prioritized.

The problem might be bigger than a time gap problem. It might actually be a complete lack of physical resources. Many of us take it for granted that there is a cavalry of equipment ready and waiting to charge to the next structure fire. Budgets, however, are limited and availability isn't always the case (fig. 11–10). In some regions of the country, mutual aid agreements aren't always solidified. What if the frontline apparatus requires maintenance and the organization can't support or afford reserve apparatus? What if our call volume demand exceeds available resources?

Fig. 11–10. An agency's budget constraints can greatly impact the availability of services. (Illustration by Fire Medic Art)

These questions should all be taken into account when developing your individualized game plan or suggested operating methods. Do not make blind assumptions about availability. Make informed and calculated decisions based on facts and real analysis. Here is an example to illustrate this point:

A standard engine company responds on a routine (be careful with that word) structure fire. Response consists of five pieces of apparatus, an assortment of engines and quints. The engine arrives on the scene to find a well involved basement fire. The officer completes a size-up and determines to initiate an aggressive interior attack off booster tank water. He radios to have the second due engine secure a water source. The officer selects a 1¾ in. hand line as his weapon of choice, flowing 150 gpm. His engine has a 500 gal booster tank.

- First do the math. How long can this crew operate in the basement on a full booster tank? *The answer is 3.33 min.*
- Next, play the what-if game. What if the second due engine has an extended response or there is a gap in coverage due to maintenance issues or normal call volume occurrence?

If this example is routine or your agency's norm, fix it! Develop operating methods that take this into consideration. Do not consistently operate like this just because it is the industry standard. Adjust procedures that add to your crew's safety and successful tactics and strategies. Don't allow that first in officer to defer water supply. Develop standard procedures for either forwarding in or reversing out supply lines by the first due engine.

Response Area Characteristics

The final variable that should be analyzed in relation to delivering tactics and assessing operational capacity is response area characteristics. The response areas in suburban America can be drastically different than those of urban cities. The landscape and all the features associated with it can have direct impacts on resource needs and operating procedures. Examples of such characteristics and features include the following:

- Water supply availability

- Building types/construction features
- Occupancy types
- Property placement
- Terrain
- Urban/wildland interface
- Street conditions (cul-de-sacs)
- Density

Water supply

It has already been discussed how the lack of a permanent water supply system can affect apparatus design. It also greatly affects tactical and procedural issues. Most urban fire departments do not have this problem. Generally they are not affected by the lack of hydrants unless there is a mechanical issue, such as a broken main or frozen hydrant. However, this is a real challenge for some suburban based agencies, specifically on their rural boundaries.

Organizations in areas without water mains or public water supply should have separate operating methods to address structural firefighting at these locations. These are much more involved than simply dropping off the large diameter supply line, wrapping the hydrant, and straight laying in. Setting up water supply without hydrants requires more manpower and resources. It also requires tactical decisions related to water conservation and precise hoseline and stream placement. Every gallon counts!

Depending on the number of responding staff on each apparatus, water supply could require the efforts of an entire engine company in a response area. This could be a problem if the first due is initiating an aggressive interior attack, but is requiring the facilitation and help from the second due. The second due cannot be expected to conduct both water supply functions and interior fire floor facilitating functions with limited staffing. When writing suggested operating methods for structure fires in water restricted areas, consider the following options:

- Apparatus to perform water supply or drafting operations
- Immediate use of booster tank water from apparatus to apparatus
- Immediate use of tank water from the first due tender (tanker) supplied directly with a supply line to the attack truck or supply truck

- Extra engines to compensate for the water supply functions
- Adequate number of tenders for desired estimated fire flows
- Extra safety concerns for interior operating crews, such as mandating a backup water supply (full booster tanks) once a permanent water supply has been established

This response scenario creates a tight window for decision making. All fireground decision making is critical; however, in an area no available water, response decisions should be precise. Time is critical and the initial judgment for an offensive or defensive attack should be made decisively. Do you realistically have the resources and water to halt fire spread and extinguish? Is it safe to enter the structure with limited water supply? If these answers are no, then accept the given circumstances and set up a defensive operation to contain the situation and protect exposures. Generally, in this application, water conservation is a major concern for long durations of flow. If it is determined that a quick interior attack can be initiated and the amount of fire can be controlled, initiate the offensive attack. For this mode, efficient application of the limited amount of initial water supply to the fire's location is critical in order to halt spread. Hit it hard and don't hold back; the time window is small. It is ineffective to attack with a reduced or insufficient flow in order to attempt to conserve water.

Building types

Like large cities and urban areas, all five building types can be found in suburban communities:

- Type I: Fire resistive
- Type II: Noncombustible
- Type III: Ordinary
- Type IV: Heavy timber
- Type V: Wood frame

One common difference between suburban and urban structures is the age of construction. Typically, newer construction is found in suburban areas due to the age of the communities and sprawling growth patterns. Some of these new construction features and techniques are not firefighter friendly:

- Truss construction
- Lightweight construction

Builders today look for the most efficient and quickest way to erect a structure. The building industry is constantly looking for cheaper building materials and methods to expedite the process from production to installation. Unfortunately, this has led to serious, detrimental consequences for the fire service.

These new manufactured structural members, such as I-beams and lightweight wood truss construction, do not fare well under fire conditions. They fail quickly and lead to structural collapse, which means interior attack times are short. Because of this, strategies and tactics must be altered to account for these disadvantages. This type of construction may dictate a greater need for resources, flow rates, ventilation, and other critical supporting functions due to quick fire spread and control issues. A series of tests conducted by Underwriters Laboratories (UL) have indicated a significant difference in structural stability in both new lightweight construction and truss materials. The failure time of a solid, non-protected framing member was 18 minutes, 30 seconds after ignition time, and the equivalent manufactured/engineered member failed in 6 minutes, 30 seconds. Preplanning and thorough size-ups are essential!

Commercial occupancies

Because the tactics for fighting fires in commercial occupancies are completely different from residential strategies and tactics, different suggested operating methods should be developed for these structures. These buildings present many more challenges to be taken in account when developing operational procedures, including a greater need for resources and staffing. There are many specific challenges and extra considerations:

- Larger areas
- Large fire loads
- Increased water supply needs in relation to large fire flow requirements
- Containment (in relation to adjacent or attached structures)
- Life hazards
- Building height
- Long hose stretches
- Ventilation
- Ventilation system control

- Elevators
- Access

Suburban fire service organizations should assess their capabilities in relation to fighting fires in their specific commercial structures. Crews must conduct preplanning and be familiar with all buildings within their response areas. Target hazards must be identified and operational action plans should be developed for each occupancy, evaluating all tactical possibilities and needs.

Property placement, terrain, and street conditions

One circumstance that shouldn't be taken for granted in relation to fireground objectives is the ability to gain access to the structure and to stretch a line to mitigate the problem. Many suburban areas have physical aspects and terrain that complicate the initiation of fireground operations. For example, residential structures that are located on one- to three-acre lots are not uncommon in some suburban regions. These homes typically are set back large distances from the roadway (fig. 11–11). In this case, the 200 ft preconnect is not sufficient to make this stretch and reach the fire. Resources required and time to assemble and place the line will be greater. If your organization is limited on staffing, do not rely on a single engine company to place this line. These considerations should be anticipated and planned for in the organization's operating procedures.

Fig. 11–11. A residential occupancy that is set back far from the street poses an additional response access challenge. (Photo by Erin McGruder)

Setbacks are just one example of landscape differences. Terrain in general can create challenges for service delivery. Narrow roadways, steep inclines, and bridge weight limitations are examples of these limitations found in typical suburban response areas. How does this affect operations, tactics, and strategies? The obvious answer is apparatus design and response time. Apparatus should be designed for or specifically detailed to these targeted response zones. Because of this access challenge, response times could possibly be delayed, which could change the entire game plan.

Street conditions also affect operations. As stated previously, suburban street systems are often challenging for apparatus access. Urban roads and streets are typically in a grid and run parallel and perpendicular to each other. Many suburban streets, generally in subdivisions, end in cul-de-sacs. The streets are also much narrower and often have tight turns and limited access. This could become a major consideration when placing apparatus. If there is only one way in, the first due apparatus could affect the entire operation with poor apparatus placement. How many times have you the seen the example of the engine stopping directly in front of the structure, restricting all aerial device access? Consider detailing an apparatus that fits its response area. Would it be appropriate to station a first due 100 ft quint apparatus in an area where it cannot turn or even gain access?

Urban/wildland interface

Another interesting dilemma that faces suburban fire agencies is the wildland interface phenomena. People move to suburbia to escape city life and to be surrounded with nature. Natural foliage, when overgrown and dried out, can become extremely flammable. Suburban fire departments are often called on to handle ground cover and natural vegetation fires. Depending on the amount of open fields and forested areas, these fires can become large and labor/resource intensive. They also can threaten all structures within their paths.

Fighting natural cover fires requires different tactics, strategies, and resource requirements. Many departments that have the potential for these types of fires equip their apparatus with specific tools:

- Special hose (forestry line or extra booster hose)
- Hand tools such as rakes, shovels, and axes
- Leaf blowers
- Forestry chainsaws
- Portable backpack water tank pumps/extinguishers

Organizations can also purchase personal protective equipment and apparatus specific for this function. These apparatus are normally smaller, 4×4 vehicles that are able to go off road or gain quick access in tight areas. A final consideration is separate policies and operating procedures in relation to fighting ground cover fires. This is a different type of firefighting, and should be treated as such.

Density

It is critical for fire chiefs and fire service organizations to monitor density patterns and trends within their communities. Population will dictate call volume and ultimately dictate resource needs. Feasibility studies should be conducted to see if the organization is effective and efficient at answering emergency response demands. These studies should determine if the agency is able to respond to emergencies within a benchmarked time period and to evaluate if engine house spacing is adequate.

Population shifts in suburban communities are quite common. It is a concern that should be observed when evaluating agency operating procedures. If apparatus is constantly unavailable due to call volume, then options should be evaluated in an attempt to overcome limitations. Also, if the later arriving apparatus has longer response times, there should be less reliance on them for support functions.

A classic example in this situation is water supply. If you are first due and realize that your normal second due company is unavailable or delayed, it is not wise or safe to defer securing water supply. In this scenario the crew should lay its own supply line. However, if you have an area that is heavily covered or engine house coverage overlaps, and apparatus arrives simultaneously, it is more effective and timely to defer water supply and go straight into attack mode.

Chapter 12

Implementation

Implementation Plan

When discussing tactics and strategies in the fire service, generally implementation of coordinated fireground actions involves enactment of a predeveloped strategic plan. Simplifying this statement, the leaders of the department evaluate all internal and external circumstances to develop operating procedures in the form of documents. These documents, depending on the terminology used by the specific agency, can be called

- Incident action plans (IAPs)
- Preferred operating methods (POMs)
- Suggested operating methods (SOMs)
- Standard operating procedures (SOPs)
- Standard operating guidelines (SOGs)

Earlier chapters discuss formal methods and systematic approaches to breaking down fireground functions, evaluating capabilities, and developing these action plans. But how do we get these operating guidelines from paper to desired actions? This process is known as *implementation*.

Implementation is actually about effecting change. This can be a complicated undertaking, which requires a plan in itself. Think about how difficult it is to create change in your organization. The fire service is a complex industry founded on tradition. Traditions give us identity and this identity drives us and gives us meaning. Its roots reach very deep and it is extremely difficult to change cultures and beliefs. However, we must adapt to our specific circumstances and create a safe and successful work environment. Firefighters must understand and perform the best fireground practices that foster an efficient, safe environment that is free from traditional bias. Tactics, as established prior, do not effectively or uniformly work for every agency. Therefore, implementation is a formal process.

Some of the best developed plans, policies, and procedures fail due to poor execution. Poor execution is a result of either a poor or nonexistent implementation plan. The development process of policies and SOGs is only half the battle and is incomplete without this implementation process. Policies should not be haphazardly proposed and disseminated to the organization without a premeditated, well thought out implementation plan.

Good implementation plans include

- Agency acceptance/buy-in for the process (organizational alignment)
- Documentation
- Plan familiarization
- Training
- Resource allocation
- Enforcement

Organizational Alignment

Implementing a plan for tactics and strategies must begin with complete organizational acceptance and compliance. Who does the procedure affect? Typically, in relation to structural firefighting, procedural implementation involves all members of the organization's suppression staff up to the fire chief. The process starts at the top and ends at the bottom of the chain; from the department chief to the newest probationary firefighter. Everyone should be on board!

From the top

It is not always effective or realistic to expect the department chief to develop every organizational action plan. Some readers might take exception to that comment. However, we must realize that heads of organizations have many responsibilities to deal with outside of the *operational* arena, especially chiefs of larger departments. Individuals actively involved in structural firefighting and affected by the circumstances that dictate direct actions must make operational decisions. These policies may or may not be made by the upper echelon of the organization; however, acceptance and enforcement by the highest power is critical for organizational alignment (fig. 12–1).

Just working on the Operational Action Plan.

Fig. 12–1. Operation action plans may not be written by the department chief, but acceptance and enforcement by the highest power is critical for organizational alignment. (Illustration by Fire Medic Art)

Operational level leaders must recognize a need for procedural implementation or change. It is also important that the organization have a formal communication system or path to inform the administration and to receive authority to initiate the process. Organizational structure must be tailored to support operational efficiency. Without formal authority, the process does not have legitimacy and will have a high probability of failure. Without acceptance and buy-in from the top, the process is powerless. Administration must be responsible for

- Matching organizational structure with strategy (giving support and resources needed for operational success)
- Creating an organizational climate conducive of change
- Creating a strategy supportive culture
- Managing human resources

Resistance to change and adverse cultural beliefs can be considered the greatest threats to strategy implementation success. Hard work and efficient leadership is required to overcome these barriers. Over time, leadership styles and theories have changed within the fire service. Much has to do with generational changes and cultural values of the individual. In relation to effecting change within an organization, it is important to evaluate what motivates the individual employee. Even though no form of leadership can be argued as the *right* and *only* effective model for the fire service, situational leadership should be analyzed. Emergency fireground directives are not the same as directives that affect non-emergency, less sensitive circumstances, and therefore should be managed differently.

Leadership styles

Leadership styles are differentiated by control and delegation of control. The leadership style and its success has much to do with who is leading and who is being lead. There are three classic leadership styles:

- Laissez faire: This style is considered largely hands-off and tends to minimize the amount of direction required. There are few controls that permit the individual to have free reign. This system works well if there are highly trained and highly motivated direct reports.
- Authoritarian or autocratic: This is the commanding style of leadership. It states, "Do as I say because I am the boss." It is based on the legitimate power of position. This system works well if direct reports are inexperienced or lack confidence.
- Democratic or participative: This style of leadership involves participation between the boss and the employee. It includes a greater equality between the leader and follower and many decisions are made by majority rule. This system works well to promote critical thinking and mutual respect within trained and motivated direct reports.

Another form of leadership, outside of classic styles, is situational leadership. This classification of leadership types is based on the work of Paul Hersey and Ken Blanchard. Through their studies, they concluded that leaders should be able move back and forth between the following four styles:

- Style 1: Telling or directing. Leaders primarily make decisions and communication is generally in one direction.
- Style 2: Selling or coaching. Leaders still make decisions; however, followers are involved and allowed to offer ideas.

- Style 3: Participating or supporting. Leaders allow followers to have an increasing say in decisions but provide coordination and guidance.
- Style 4: Delegating. Leaders allow capable employees to perform largely on their own and make their own decisions.

Situational leadership allows for a more flexible and balanced leadership. The leadership style varies dependent on the needs of the subordinate and the situation. This situational style is more applicable to the fire service due to the diversity of circumstances and vast differences in experience levels of employees (fig. 12–2).

Fig. 12–2. Supporting leadership coordinates and guides followers with an increasing say in decisions. (Photo by First Alarm Production Fire Ground Photography)

Example: Would you allow two probationary firefighters to operate independently without direction or with minimal direction and the ability to make fireground decisions? The answer should be an obvious "no." The situation of an emergency fireground scene and the needs of the followers dictate a directive style of leadership.

Now use the same example with a 20-year veteran who has been ordered to conduct outside ventilation independently from his officer. This firefighter doesn't need his hand held and can make safe decisions about ventilation practices without direct commands. The firefighter can make formulated opinions; however, is still under the direction of the officer.

It should be noted that a completely hands-off style of leadership is not appropriate on the fireground. The circumstances on the fireground do not allow for committee decisions and orders should be followed in relation to the chain of command.

Successful leaders are those who are able to effect change so that employees want to complete tasks. In order for this phenomenon to transpire, leaders must possess the values that followers identify with and hold with equal value. Examples of these virtues include

- Courage
- Honesty
- Integrity
- Excellence
- Respect
- Loyalty
- Faith

Resistance to change

As stated earlier, implementation involves change. Change may be subtle or it might be a major alteration in service delivery. Regardless, resistance to change should be anticipated and strategies should be developed to combat this resistance. There are three commonly used strategies for implementing:

- Forced change strategy: This strategy involves giving orders and enforcing orders. This is a fast solution; however, it involves a low commitment and higher resistance.
- Educative change strategy: This strategy presents information to convince people of the need for change.
- Rational or self-interest change strategy: This strategy attempts to convince individuals that the change is to their personal advantage.

Culture change

A major underlying theme in this book is changing behavior and attitudes toward suburban fire tactics and operations. As stated earlier, it doesn't require a large study to realize that suburban fire agencies are different from their urban counterparts. Yet many suburban fire agencies still operate as if they were large urban fire services. Much of this is due to resistance to change, and much more has to do with ego. The ego has to be removed from operations. It is a vehicle that will lead to our demise. The saying goes, "Progress impeded by 250 years of tradition." Suburban fire tactics are by necessity challenged by traditional operations. For the safety of fire agencies operating in suburban America, we must alter our tactics to reflect our situation and circumstances. Unfortunately, in many cases this is drastically different than urban firefighting.

Implementation often requires a large change in attitude and beliefs. This may be the single largest hurdle to the implementation process. It requires strong leaders and an emphasis toward a complete cultural change.

Organizational accountability

The implementation process begins at the top with the chief and then works its way down the chain of command. All members of the agency should understand their roles in relation to the application of set procedures. Procedures are not to be picked and chosen from. Implemented procedures are not optional and their success is determined by complete staff compliance. This complete compliance can only be accomplished through a detailed process of steps:

- Communication
- Thorough training
- Evaluation
- Reevaluation

In order to assure accountability, milestones should be set and mechanisms should be in place to monitor progress toward these milestones. Members should see how their individual performance tangibly assists the organization and adds value.

In relation to structural firefighting tactics and strategies, all suppression staff members should understand their roles and easily see the benefits of effectively contributing. For example, chief officers should see the value of directing the optimal fire attack and evaluating operational effectiveness. Company officers should be able to determine functional priority in relation to the desired suggested operating guidelines and commit their companies to those selected functions. Firefighters should understand the objectives of the selection functions and conduct those functions in a disciplined and competent manner. Each member on the fireground has a purpose and role, and should add value toward the successful completion of strategic objectives.

Once every member in the organization supports the implementation and upholds individual accountability, the organization has been aligned. Cultural attitudes and organizational change may be just as important as operational awareness. If this alignment can be achieved, implementation will have a greater chance for success.

Timelines

Implementation plans should have an assigned timeline to assist with project accountability. All objectives should be SMART:

- Specific
- Measurable
- Acceptable/achievable
- Realistic
- Timely

Timeliness is a critical factor in the success of any project. Without an assigned timeline, it is unrealistic to expect that delegated tasks will be completed. Each task within the implementation plan must have an expected completion period.

Each assigned task must be grouped and categorized in relation to the project's critical path. The critical path designates relationships between tasks and their required predecessors. Simply stated, some tasks are dependent on others and cannot be initiated until these tasks are completed. For example, if you were implementing a new suggested operating guideline for technical rescue, it would be critical to train selected operating members to a proficiency level (technician) and obtain the required rescue equipment before initiating high risk operations. These dependencies are further illustrated the critical path method charts later in this chapter.

Finally, when constructing timelines, it is also necessary to include milestone markers. Milestones indicate when critical elements are to be completed and indicate benchmarks within the project. These markers assist with breaking up the project into attainable and manageable periods and provide a framework for organization through a stepping stones approach.

Timeline tools

There are many tools developed for timeline implementation and project management. Three such examples are easy to follow and applicable for fire service project management:

- Program evaluation and review technique (PERT) chart
- Critical path method chart
- Gantt chart

Program evaluation and review technique (PERT) chart

The PERT chart is a project management tool for determining in advance the amount of time required to complete a project. On this chart, each step of the process is assigned an estimate for the time of completion according to a best, worst, and most likely scenario estimate. Once these estimates have been concluded, an average completion time is then determined.

PERT charts have typical elements:

- Numbered rectangles are nodes that represent milestones
- Directional arrows represent dependent tasks that must be completed in a sequence
- Diverging arrow directions indicate possible concurrent tasks
- Dotted lines indicate dependent tasks that do not require resources

Critical path method chart

A critical path method chart indicates and defines the longest sequence of dependent tasks that leads to the completion of the project. This type of chart draws attention to the critical path and illustrates that any delay in a task along the critical path will cause a delay in the overall project. It allows for the identification of tasks along the critical path that will need to be given special attention. The critical path from beginning to completion is indicated.

Gantt chart

A Gantt chart is a matrix that lists on a vertical axis all the tasks to be performed. Each row is designated with an identified task. The horizontal axis consists of columns that represent estimated task duration, resources, the name of the person assigned to complete the task, and one column for each period in the project's duration. Linking dependent tasks can indicate critical paths.

Documentation

The documentation process should begin long before the implementation phase of the project is initiated; however, it is an essential element in the implementation plan. Documentation should start at the infancy of the project and be continuous throughout. In relation to applying newly developed suburban fire tactics and strategies to daily operations, documentation begins with noting capabilities and existent operations practices. This includes documenting what is working and what is deficient. It includes all correspondence, requests for the authority, and responsibilities for the tactical implementation. The process should archive all major events and findings that focus on operational improvement. It should be conducted with the intent to show legitimacy and purpose, while fostering a template for continuous operational improvement.

Documentation should be the medium to transfer all major tactical and decisive expectations throughout the organization. It is a tool that organizes all planning decisions and gives a clear and concise template for expected actions. In relation to implementation, we can distinguish between the documentation of the actual operational action plan and the documentation of the entire process (from planning, through implementation, to reevaluation).

Documentation for the process

As previously stated, archiving all findings and events is critical for success. It shows the path from start to finish and will hopefully answer the question of how and why decisions were made. Items that should be documented are

- Authority for initiation of the project or change in procedures
- Definition of the problem
- A SWOT analysis for the organization (strengths, weaknesses, opportunities, and threats)
- Statistical data supporting decisions (e.g., response data)
- Documentation of planning meetings
- The developed action plan (suggested operating guideline)
- The training process
- Evaluation (postincident critiques and statistical data)

Documentation of the action plan

The goal of this entire process is to produce an effective action plan that satisfies the needs of the organization on a consistent basis (fig. 12–3). This can only be achieved with a written plan. The actual document should be clear, concise, and serve as a template for tactics and strategies. It should be a road map for implementation. As stated in the beginning of the chapter, your organization may use a different term for the action plan, but regardless of the terminology used its intent is the same: to provide sound, safe operating procedures on a consistent basis.

Action Plan Documentation

The actual document should be clear, concise and serve as a template for tactics and strategies.

Fig. 12–3. The key elements of an action plan should convey all expectations needed for consistent implementation. (Illustration by Fire Medic Art)

Chapter 2 provides the methodology for creating an action plan document. For implementation purposes, key elements should be present within the document to assist with achieving desired operational results:

- Purpose (to understand prioritization of tasks)
- Preferred task assignments
- General statements (accepted operating principles)

These key elements should convey all expectations needed for a consistent implementation. Once this document has been developed, it should be made accessible to all members of the operating staff. All subsequent changes should be communicated in a formal communication process, usually a written memo format.

Finally, don't be afraid to alter the document. Not all plans work the first time. During the implementation phase of the process, problems may occur that will require alteration or contingency plans/strategies to be implemented. Often this occurs because variables are either overestimated or underestimated.

Contingency plans must be developed. They are the plan "B" if something goes wrong. In the event of uncontrollable or unforeseen circumstances, they can be implemented in the place of initial assumptions. Contingency plan development is a proactive rather than reactive approach to project risk management. It can save the implementation plan if the unforeseen occurs.

An example of a contingency implementation can be as simple as this example:

> *After considering apparatus placement, booster tank size, and staffing, it has been determined that the best operational practice is for the initial arriving apparatus to not defer securing a permanent water source. The initial suggested operating guideline details that the initial engine will utilize booster tank water for initial fire attack and allow the close secondary engine to lay a supply line and secure the water source from the hydrant. Unfortunately, during implementation and months of documenting after action reports, it has been indicated that the first arriving engine ran out of water on numerous occasions prior to the arrival of the second due engine.*

In this scenario the arrival time of the second due engine was grossly miscalculated. However, during the planning phase of the SOG development, this variable was considered and an alternative was developed. The alternative strategy established the need for the first due to either forward lay or reverse out its own supply line to the hydrant. Implementation can now resume with a small setback, as opposed to complete failure over a simple operational deficiency.

Familiarization and Training

The next phase in the implementation plan is the training component. Depending on the type of implementation change, this phase can take many forms. Regardless of whether this is an implementation a new procedure or the enforcement of current /standing procedures, the training division is an essential component of any fire service organization. Without the training component, the implementation process cannot be initiated.

In relation to the fire service, there are certain considerations and conditions that require or indicate training needs:

- Legally mandated training
- Performance during emergency operations
- Annual refresher or recertification requirements
- Post incident evaluation of events/deficiency identification or corrective training

- Changes in operational procedures
- Implementation of new policies, procedures, or equipment
- New service delivery
- Preparation for advancement
- New personnel duties/assignments

Fig. 12–4. Without the training component, the implementation process cannot be initiated. (Photo by Steve Heidbreder)

The training that can be delivered to satisfy these needs, deficiencies, or improvements can take many forms. The delivery methods that should be considered depend on the type of knowledge, skills, or abilities that need to be acquired or learned.

Learning, by definition, is

1. The act, process, or experience of gaining knowledge or skill
2. Knowledge or skill gained through study
3. Behavioral modification through experience or conditioning

Although the forms of fire service training vary widely, there are four common methods:

- Presentations
- Discussions
- Demonstrations
- Practical training evolutions

When developing a training program, the training officer should begin with objectives. The objectives should consistently answer the need for the intended audience's desired knowledge, skill, and ability improvement or acquisition. These objectives should be categorized into the three domains of learning:

- **Cognitive:** Comprehending information, organizing ideas, analyzing and synthesizing data, and applying knowledge.
- **Psychomotor:** Learning by relating movements or muscular activity with associated mental processes.
- **Affective:** The acquisition of behaviors involved in expressing feelings in attitudes, appreciations, and values.

It should be noted that many learning objectives are not mutually exclusive in relation to domain category. Some objectives may fall within more than one category.

Within the cognitive domain there are six levels of learning:

- **Knowledge:** define, repeat, record, list, recall, name, describe, identify, match, select, outline, write
- **Comprehension:** classify, explain, summarize, convert, predict, distinguish between
- **Application:** demonstrate, compute, solve, modify, arrange, operate, relate, adapt, apply, implement
- **Analysis:** differentiate, diagram, estimate, separate, infer, order, subdivide
- **Synthesis:** combine, create, formulate, design, compose, construct, rearrange, revise
- **Evaluation:** judge, critique, compare, justify, conclude, discriminate, support

As an example, we specifically target the development of a training program designed for the implementation of a newly developed suggested operating guideline for structure fires.

Start with SMART (specific, measurable, achievable, realistic, and timely) learning objectives:

1. *All operational staff members will be familiar with, explain, and understand the newly developed suggested operating guideline with 100% accuracy in the classroom, prior to practical application.*

Given this learning objective, we can identify the domain of learning as cognitive. The cognitive level is very basic and lower level: knowledge and comprehension. The best method of training delivery in this example is a formal presentation. A presentation would accurately provide all information for this comprehension.

2. *All operational staff members will be able to understand and evaluate the importance, relevance, and effectiveness of the new suggested operating guideline in order to assist with operational ownership and acceptance and comply 100% with set mandates.*

This learning objective falls within the cognitive and affective domain. This objective is attempting to affect attitudes, appreciations, and values. This is a very progressive learning strategy. Many firefighters and fire officers from the past will argue that important policies, such as operational procedures, are not up for debate and should be mandated. Keep in mind that adults learn differently and variations can be based on different cultural backgrounds. If the trainer/leader can effectively gain support through attitude and behavioral modification, the objective will have a much greater chance at success and compliance. This is very true if it is a policy change to a procedure that has existed for many years, or a policy that is a cultural change. For operational procedural changes, this can often be the case.

For this example, the best teaching technique is probably the led discussion. It will require an experienced training officer who has extensive knowledge and credibility in order to affect attitude.

3. *All operational staff members will be able to demonstrate and apply the new suggested operating guideline with 100% accuracy given a structure fire scenario.*

This objective is within the cognitive domain; however, is past the comprehension level. It falls within the application learning stage. A presentation and discussion is good to begin the knowledge and comprehension of the material, but for application a demonstration and practical evolution is more effective. For this objective, the best solution may be a *tabletop* exercise. This is a scenario-based exercise that works well in a classroom without the hazards of the fireground (fig. 12–5).

Fig. 12–5. Tabletop scenarios fall in the application level of learning to solidify concepts and reinforce global awareness. (Photo by Erin McGruder)

4. *All operational staff members will be able to demonstrate the new suggested operating guideline and conduct all associated assigned tasks with 100% accuracy during a live fire training scenario.*

This objective is both cognitive and psychomotor. There is only one way to effectively produce this learning objective and that is through practical application. Live fire training is the best way to prepare for this implementation; however, your organization must be 100% compliant with all live fire training regulations set by NFPA 1403. Live fire training can be dangerous and all safety considerations and procedures must be followed without exception. Also, to be effective, the training must be as realistic (within safe means) as possible. "Practice like you play" means to do everything as if it were the real deal. The learned psychomotor skills will be remembered later during the real event.

A developed training program for implementing operating procedures, specifically structure fire operating methods, preferably should be conducted in phases:

- Classroom presentation (familiarization)
- Tabletop scenarios (application)
- Live fire training (application and psychomotor)
- Company level training follow-up

Reinforce training through company training. Company level training is an opportunity to fine tune fireground functions. It also gives the opportunity to train on a smaller, individualized basis that can be more beneficial for specialized learning. Company level training can prove to be a great team builder for a crew and build stronger knowledge, skills, and abilities.

Promotional implementation

Implementing tactics and strategies should be a complete organizational endeavor. It shouldn't be isolated to the training component. It should be implemented at every possible arena of the organization. A prime opportunity is incorporating developed operating methods into promotional assessment centers. This can easily be accomplished through applying concepts and methodology directly related to the implemented operating procedures to training events:

- Written testing
- Oral reviews
- Tabletop scenarios
- Practical fire scenarios

Implementing the suggested operating guideline into the promotional process gives the document legitimacy and reinforces its meaning and purpose. It also guarantees that officers will have strong proficiency for accepted tactics and strategies.

Resource Allocation

Resource deficiencies might become apparent during the planning and development phase. The implementation of operational procedures may require additional allocations. In extreme situations, entire engine houses and companies will have to be created to ensure operational effectiveness. For many organizations, this is a complicated and impractical undertaking. However, if it is an absolute necessity and community demands justify the expansion, there are means to accomplish this objective. The biggest hurdle is the ability to finance the undertaking. There are three finance vehicles an agency can use (depending on its political structure) to achieve the goal:

- Income through bond issuance

- Income through a tax increase
- Additional income streams from billing of services (e.g., ambulance billing)

Each of these three finance vehicles requires effort and a planning component. It is extremely unpopular to charge for services and to increase taxes. It may require a major marketing campaign and public relations strategy. All will require time, money, and effort.

If expansion is not an option, reallocating essential resources may have to be accomplished more creatively. One of the greatest examples of this allocation was the development of the quint concept.

Chapter 11 gives as an example of combining resources the quint fire apparatus, which provides the answer to staffing an engine and truck with one vehicle. This apparatus meets the needs of tight district and city budgets by eliminating or reducing the need staff to an apparatus that has limited versatility.

The quint concept has been embraced by many different fire agencies across the country, including many urban. It is not the answer for every agency, however. Certain agencies still operate with a traditional system of divided functions for separate trucks and engines. Typically, many of these agencies do not have the resources for one truck for every engine. For this example, the agency must find the optimal placement for their truck(s). Call frequency and population trends are a good start in determining this allocation. The key is to not delay truck company functions and keep them coordinated with the fire attack.

Other instances of reallocated resources include the combining of companies for efficiency. Many suburban agencies have turned to these solutions to solve budgetary shortfalls. Chapter 11 also illustrates some creative options for solving budgetary constraints:

- Rescue/engines
- Pumper/tankers

Allocation example

To illustrate this point, allow me to share an example from my own organization. My suburban organization had operated a fully staffed heavy rescue truck through the early 2000s. Unfortunately, with the economic downturn experienced throughout the United States after 2001, our district began to realize a decreased budget. After a failed attempt at a tax increase, the leaders of the organization realized that the only answer was to eliminate one company. The only company that made sense at the time was the heavy rescue, which already shared an engine house with a quint. To accomplish this budgetary strategy, this company was combined with a centrally located engine and a rescue/engine was established. This company operates as an engine but is specially detailed for all technical rescues.

Combat ready/operational readiness

The new buzz phrase in the fire community is to be *combat ready* (fig. 12–6). This term can take on a variety of meanings and themes. The major underlying concept of combat ready is operational readiness. Yes, your organization can do the research, properly place apparatus, develop the perfect SOGs, and be compliant with staffing standards, yet still fail! The reason is because they are not operationally ready for the job.

Fig. 12–6. Combat readiness is a state in which operations personnel are fully prepared to perform job tasks. (Illustration by Fire Medic Art)

What are components of combat readiness? Combat readiness encompasses more than one variable. The concept includes everything that enables operations personnel to be fully prepared to perform job tasks. Examples include

- Training

- Apparatus
- Equipment
- Physical fitness of staff members
- Mental preparedness

For operational success, it is essential that all staff members be prepared to perform their assigned job functions for effectiveness and efficiency. The most obvious and visible component of this is training. Firefighting training should be an everyday event in addition to training received during implementation. Because this job entails many high-hazard, high-impact events, skills need to be refined and sharp at all times.

When discussing the human resource variable in detail, we analyze individual knowledge, skills, and abilities (KSAs). Training is only one factor in acquiring, retaining, and maintaining KSAs. In the profession of firefighting, being prepared also requires physical and mental readiness.

A firefighter's physical capacity is just as important as the working capacity of the responding apparatus. The physical fitness of a firefighter is not only important to complete assigned fireground tasks, but also important for firefighter safety. This safety involves the individual and all co-workers who might have to rely on the individual in life or death situations. In 2010, firefighter heart attacks accounted for 56.4% of all line-of-duty deaths in the United States (48 of 85). This was the leading nature of all fatal injury statistics. Not getting involved in a physical fitness program is selfish and just as hazardous as not training for job specific tasks.

Physical fitness programs should include specific processes:

- A physical fitness committee established by the organization
- Yearly medical evaluations
- Individualized physical fitness programs for all employees, including standards and goals
- A training and education component
- Physical fitness program maintenance (an established time for employees to engage in physical fitness activities during the work period)
- A follow up, retest, and documentation component
- Dietary consultation

Wellness programs should be established to account for personal choices:

- Nutrition

- Hypertension
- Cessation of tobacco use
- Weight control
- Physical conditioning

Fire service organizations should also be concerned with the mental well-being of their members. Decisions made on the fireground have to be made quickly and be clear and concise. This requires the firefighter to be at a state of complete mental readiness at all times. Due to the nature of this profession, we often see situations and experience circumstances that may cause stress and mental anguish. Therefore, it is the responsibility of the agency to establish programs that can help their personnel, such as the following:

- Member assistance programs, to account for
    - Substance abuse
    - Stress
    - Personal and interpersonal problems
    - Depression
    - Anxiety
    - Divorce
    - Financial problems
- Critical incident stress management program

Apparatus considerations

Implementing tactics and strategies also relies on the operational readiness of equipment and apparatus. Simply stated, operational staff members cannot conduct fireground functions efficiently or effectively without the right tools. Do you have the right firefighting tools for the job you are trying to accomplish? Are your tools accessible? Is your apparatus ergonomic and does your organization promote safe operating practices?

The right tools. For this section, the term tool is an all-inclusive term. Do not assume that we are merely discussing hand tools. For the purpose of operational readiness, consider the following tools and examples:

- Forcible entry tools
    - If you have high security considerations with a large presence of steel bars and grates, it is essential to have a forcible entry/cutoff saw.

- If you have a large population of commercial structures, it would be appropriate to carry through-the-lock tools.
- Position all forcible entry equipment for immediate use (accessible and close to the assigned firefighter).

- Hose tools and appliances
  - If you have procedures established for stretches or lays that require the use of hose tools or appliances, considerations should be made for easy/quick access. For example, for long hand line stretches, it might be a common policy to deploy a 3 in. line to a gated wye and then attach 1¾ in. hand lines to the wye. A system should be set in place for this easy/rapid deployment, including preattached appliances and prebundled attack line.
  - Another example of quick access for hose appliance is the hydrant bag attached to the end of the large diameter supply line. If it is your policy to straight lay into the fire scene, the hydrant bag (containing hydrant wrenches and specific adaptors) makes the water supply connection flawless and quick.
  - Have all necessary adaptors and appliances stored close to the pump panel. Create an engineer's compartment.
- Hose and hose loads
  - Have multiple, preconnected hand lines of adequate and different lengths (determined by area familiarization and preplanning).
  - Prepare attack line storage/loads suitable for minimal staffing deployment.
  - Include different sized attack lines for different flow options (include a 2½ in. preconnect for large water application options).
  - Load your hose bed consistent with your response area (lengths) and load hose consistent with policy and procedures (forward load for straight lays, and reverse loads for reverse outs).
  - Have supply hose set up and accessible for quick, truck-to-truck connection.
- Ladders
  - Ladders should be able to be deployed by minimal staff members extremely rapidly.
  - Tie the lanyard for quick raises.

- Nozzles
    - Nozzles should be consistent with organizational procedures and practices, determined by the desired use; either a straight stream or fog pattern (attack method).
    - Consider nozzle reaction (fog nozzle) if operating with minimal staffing.
    - The nozzle should be able to deliver a variation of desired amounts (gallons/minute).
- Master stream devices
    - Have an option for quick deployment of a master stream device (deck guns, elevated ladder pipes, monitors).
- Miscellaneous tools specific to response area
    - Does your response area have any special considerations? For example, do you respond to large commercial areas that may require a foam application? If so, extra foam and foam application appliances should be considered for the apparatus.

Tool and position assignment

Fireground functions cannot be completed without the specific position assigned to complete them with the proper tools. Often, implementation failure of tactics and objectives can be pinpointed to this aspect. If crews report to the fireground without hand tools or assigned equipment, jobs cannot be initiated effectively or quickly. Fireground success often relies on the swift and rapid action of competent firefighters.

At the beginning of every shift, assignments should be made by the office and communicated to the assigned staff. These assigned positions should be consistent with capabilities and current operating procedures. They should be made with consideration for completing individual fireground functions and satisfy both truck and engine company operations.

Engine company assignments must be able to satisfy line placement functions. Examples of line placement positions include:

- Nozzle
- Backup
- Door
- Control
- Officer

Truck (facilitating) assignments should be able to satisfy any determined facilitating function. Examples of truck positions include

- Forcible entry (irons)
- Outside vent (hook)
- Can (extinguisher)
- Officer (thermal imager/hand tool)
- Roof (hand tool/saw)

Enforcement

Finally, no implementation plan can be complete without strict enforcement. Policies can be developed, but they will not be followed without some degree of enforcement. It is up to the leaders to enforce all procedures and create quality control measures to ensure successful implementation. Enforcement of policies can require different measures of latitude.

It is human nature to be uncomfortable with conflict. Conflict is what creates change but it can also be a barrier to implementation. Policy manuals should be clear on the consequences for undesired actions and consistent with discipline delivery. In the fire service, this delivery typically takes the form of progressive discipline. No one likes to be disciplined or be the bad guy; however, policies and procedures are more important than individual beliefs and attitudes. Attitudes must be shaped to correspond with implementation and organizational alignment. Policies and procedures must be followed without bias or random compliance, and there should be no tolerance for deviation. To succeed, enforcement parameters must be in place before the implementation plan is initiated.

Reasons for Failure

Implementation plans can fail for various reasons. Discussing why plans fail can better help us avoid implementation disaster.

The following is a checklist for disaster:

- No support from the top

- No authority to implement
- No support or compliance from operational members
- Poor documentation
- No communication within the organization
- Under- or overestimation of organizational strengths and weaknesses (biased situational awareness)
- No developed contingency plans
- No accountability
- No timeline
- Cultural bias/inability to change
- Insufficient training/familiarization
- Poor leadership
- Insufficient resource allocation
- Lack of readiness/preparedness of equipment or staffing

Implementation of the Game Plan

The implementation of plans, policies/procedures, and strategies can be a challenge. If not done correctly, the entire process will fail. It is a fragile element that must be conducted with precision. It cannot be thrown into the mix haphazardly.

This chapter suggested numerous variables that play key roles in the implementation phase. These variables represented human resource elements, physical resources, and specific management principles.

The process requires a strong leader who will be required to be a change manager and a visionary who is able to hold the organizational accountable for the plan. The organization should be prepared for:

- Organizational alignment
- Developing and fostering effective leadership styles

- Change
- Developing organizational accountability from the top to the bottom of the organization
- Developing timelines
- Familiarization and training
- Developing promotional implementation
- Resource allocation
- Refining operational readiness
- Developing procedures/policies for enforcement

Remember, begin implementation with a *plan to plan* and develop a keen sense of situational awareness that is fair and unbiased. Most importantly, prepare for change. Change is often a good thing.

Chapter 13

Reevaluation and Future Planning

This chapter was contributed by Captain Michael Digman of the Metro West Fire Protection District, St. Louis County, Missouri. Mike serves as the standards of coverage manager for the district's accreditation process, and is responsible for the evaluation, review, analysis, and interpretation of the district's response performance data.

Today's emergency service agencies are tasked with an exponentially increasing list of duties and responsibilities without the corresponding and compulsory increase in funding. In today's fire service, increasing demands for service are usually paired with corresponding *reductions* in budget, rising operational costs, and an increased cost of replacement for apparatus and equipment. In essence, we are being asked to do so much more with little or no increase in funding.

Firefighting has transitioned to an all-hazards response. Agencies that were solely responsible for fire suppression are now also tasked with technical rescue, hazardous materials response, water rescue, public assistance, and emergency medical services (EMS). Throughout the country, EMS has become the norm rather than the exception. The national average is that well over half of the requests of fire departments are for EMS services. Simply stated, today's fire service is busier than ever before. While the instance of structure fires is trending down across the country, the annual alarm totals continue to rise.

The fire service has traditionally been a *reactive* profession. That is to say, we do not pre-position hose lines, apparatus, and personnel *and then* the house catches on fire. From the moment that the first wisps of smoke fill the air, vehicles slide out of control, or a patient begins to become short of breath, our crews are behind in the fight. We, as a profession, have mastered the art of quickly leveling the field upon arrival and working to get the upper hand on the emergency as it evolves. The question then is, "How do we prepare for the increased call load of varying hazard types with the budgets and constraints that are constant within the organization?" The easy answer is with a thorough, detailed, honest evaluation of resource deployment and historical alarm data trending, with an equally honest and open reception of the data and what it will ultimately recommend.

The facts evaluated and data compiled will paint a no-nonsense picture of what level of service the agency is currently providing. These facts are indisputable; they are derived from the alarm reporting recorded by the fire agency. They will help guide your future organizational decisions.

The numbers over a sampling of years will paint a very vivid picture of the effectiveness of your organization. In this chapter we define turnout times, response times, resource allocation, effective response force, and critical task analysis. All of this information is readily available to you as you analyze your department's data to make the appropriate decisions for the future.

In the *reevaluation and planning* phase of applying suburban fire tactics to your organization, it is imperative that you take a close look at your jurisdiction, present response package, response trends, and future planning. Without this honest and comprehensive assessment, there is no way to definitively state that you are providing the best service possible to the citizens you are sworn to protect.

Be prepared to take a tough look at your agency as this data unfolds. It is possible that the data will tell you that part of your population is underserved. It may say that you need to add assets to a company's response area to meet your goals. It may recommend more firefighters be hired or a fire station be relocated. Ultimately, it is up to the governance of the agency to decide how to interpret this information and best direct the organization's goals for the future. Without this analysis, however, there is no honest way to say that you are protecting your citizens to the best of your abilities.

While we are not advocating that you relocate companies or engine houses simply because of the information that is produced from the evaluation process, you may discover that, in fact, the call volume for Station 4's ambulance necessitates an additional rig and staffing and that your heavy rescue may better serve the public in a more central location. You may determine that in order to accomplish organizationally identified critical tasks, the addition of an additional firefighter is needed on a particular engine or all the engines within the agency. We are, however, suggesting that you take the time to interpret the data that you compile every day through daily fire station reporting to better your organization.

Examination of the Current State of the Department

It sounds like a no-brainer, but do we know exactly what we have, where it is, and whether or not it is positioned in the best place to maximize the efficiency and benefits of the unit?

As we've identified, the fire service is an industry steeped in history and tradition. As communities grew during the post-war suburban expansion, single station fire departments found the need to add additional staffing, companies, and eventually more stations to cover the areas that they were obligated to protect. Our predecessors, to the best of their abilities, positioned stations to serve the booming expansions of the time. Grabbing real estate as it became available or was apportioned to them by the city, firehouses sprang up across the country as the suburban housing expansion progressed.

Much of this expansion was dependent on the socioeconomic development of the area and followed the developments as they grew within the jurisdiction. New communities tend to develop in clusters, and the undeveloped area between larger developments fills in later. Fast forward 50 years and what was once farmland is now subdivisions, multi-family complexes, commercial and retail developments, assisted living communities, and schools. Stations placed to serve the population centers of the agency in 1965 may not meet the needs of today's community. In fact, they may be located in an area that is *detrimental* for the area that they are required to serve.

Every major metropolitan fire department has experienced these growing pains and subsequent reevaluation of asset deployments. Engine houses get shuttered, relocated, and combined with other companies as the first due areas they serve change with the times. Firemen are not fortune tellers. There is no way to determine with 100% certainty that a station that was centrally located when it was built will still be effectively located in 30 years. What is certain is that with proper evaluation and analysis of the data that is already being compiled by the agency during every shift, a plan to better serve the public into the future can be developed.

Physical inventory

The physical inventory is the easiest task to complete while conducting an evaluation and future planning process. It is simply that—an inventory or catalog of what is available to the organization for mitigation of emergencies.

While it may seem to be an overly simple process, it is critical that you first understand what exactly you have to work with in order to properly evaluate your organization. While larger capital holdings like fire stations and ladder trucks may seem obvious, taking the time to compile a good, thorough resource list that answers the following questions is helpful to the future planning and evaluation process.

- What do we have?
- Where is it located?
- Where can we get a resource if we don't have one available?

What do we have? When performing the inventory, it is important to look closely at what the organization currently owns in both fixed and disposable assets. The make, model, and year of manufacture of engines, ladders, brush units, and water rescue vehicles will allow the agency to prioritize a replacement schedule. This is a good time to look at the physical inventories of the units and develop a comprehensive list of the cache of equipment that each unit brings to the scene. Additionally, identifying an inventory of perishable items such as Class A foam or stockpiles of reserve air cylinders is important so that, in the event of a large incident, officials know from where to draw resources.

There are some things to consider when evaluating your equipment:

- Apparatus type (ladder, engine, tanker/tender)
- Booster tank amounts
- Hose bed configuration
- Mileage/hours on apparatus
- Grade (replace cost versus worth)
- Aerial length
- Apparatus heights and lengths
- Rescue tools or other specialized equipment

Where are our assets located? Perhaps more important than what we have is where in our organization the equipment is located. A fire officer on the scene needs to know where the proper tools to mitigate the emergency are within the agency and how to request assets in a timely fashion. It is critical to the outcome of the emergency that he or she knows the rough travel time of an additional ladder, how long to expect to wait on a water tanker to get to the scene, or how soon to call an upgrade to a second alarm in order for that equipment to be effective.

A while back, during a phase of construction within a station in our area that required movement of apparatus, we joked that, "We are the only organization that has a 100 ft platform in the country and a brush rig in the city." Obviously, these equipment assignments no longer made sense for our agency and the apparatus was relocated at the completion of the project. In the interim, command staff had to adjust the detailed equipment to the box assignment and bring in companies that might not normally respond to an address in order to make sure the closest appropriate assets arrived where they were needed.

That situation was temporary. Decisions to locate a station or piece of equipment in an area might meet the current needs of the organization, but may fail to do so at a later point when the community continues to grow and sprawl.

As previously discussed, most suburban fire agencies began as a single engine company, and the expansion of the departments was largely driven by population booms and increases in requests for service. Appropriately, early organizational leaders positioned stations in these areas as they expanded.

Later in this chapter we discuss geo-plotting alarms to identify trends in service request density. By analyzing and interpreting information you can learn if your stations and apparatus are in the best locations; compile necessary data to justify requests for hiring, budget increases, and equipment purchases; and use the data to formulate strategies to improve infrastructure deficiencies.

Through conducting an inventory of what is actually on the street, and what street that equipment is on, organizational leadership can identify and make financial plans for the demands placed on their organization by the public.

Where can we get a resource if we don't have one available? An evaluation of your mutual aid companies may also assist in this process. Does our organization need to prioritize the purchase of a 100 ft platform, and by doing so forgo the much needed station renovation? What if the neighboring jurisdiction has a platform and there is a reciprocal aid agreement that comes to your emergencies in a timely fashion? Many smaller communities that cannot afford heavy rescue, water rescue, and ladder tower units in their own jurisdiction will contract a cooperative agreement to jointly purchase equipment with neighboring agencies. Others will each buy one asset and make their equipment available to the other organizations in an automatic mutual aid agreement (fig. 13–1).

However your agency buys, locates, and staffs equipment, make sure that you, your officers, and your administration are well versed on the location and inventories of the units detailed in on your run.

Fig. 13–1. Mutual aid can provide much-needed supplemental resources. (Photo by Thomas Vatterott, Jr., of Metro West Fire Protection District)

Many alarm centers will develop and maintain an apparatus catalogue to keep up with trends in the area, as well as the changes that occur in apparatus and equipment inventory. This resource may be helpful for the evaluation of your agency's needs. Remember that other municipal, state, and federal assets can be added to your catalogue through proper requests.

Geographic profiling

When conducting an assessment of the agency, it is imperative that the agency looks within itself to find out exactly what is contained within the organization's borders. While it seems elementary to discuss the topographical composition of the area or dwell on a major highway in the middle of the jurisdiction, these very factors affect the responses that we are called to mitigate. In essence, the composition of the response area determines what type of alarms the emergency services agency must be prepared to handle. Take the time to identify these elements:

- Major highways/interstate highways
- Residential developments
- Educational complexes

- Infrastructure (roads, water, bridges)
- Commercial/industrial process areas
- Rivers, lakes, and other bodies of water
- State/federal parks
- Tunnels
- Ferries
- Rail yards/railways
- LPG/CNG or other pipelines
- Remote access areas
- Airports
- Hazardous materials transmission routes
- Hazardous materials production facilities
- Nearest mutual aid resources

These factors all should be identified and a corresponding assessment should be made to determine if the agency is equipped and prepared to handle an emergency in these areas. Does your department own or have an agreement with a vehicle that can extricate an injured hiker from the middle of the state park? If there were a boat fire or emergency on a lake, could you get to the scene with the proper equipment in a timely fashion?

NFPA (National Fire Protection Association) response time benchmarks are guideline recommendations of how long an agency should take to deliver equipment and personnel to a scene, as well as initiate suppression, conduct ventilation, and perform other required tasks. These times vary as a result of several factors, but the idea of meeting a goal or benchmark time is an important theme that organizational leaders need to begin thinking about. Nothing is worse for a fire company than losing a citizen to fire, burning a house to the ground, or taking bad press from taking too long to get to the scene. Each second that passes makes the structure or emergency that much worse for us as responders as well. Understanding which areas have this potential helps leaders when developing strategies to meet the needs in those areas.

The infrastructure of your agency plays a critical role in the services that can be delivered to your citizens. Is the main east-west route in your agency a wide multi-lane road that easily accommodates responding apparatus, or is it a narrow two-lane road? Are there direct routes to access your entire jurisdiction or is substantial off-tracking required by the first due companies to reach a scene? Does the infrastructure in part of your jurisdiction prohibit any of your apparatus from making it to the scene because of bridge clearances, weight restrictions, or other size issues? Is there a part of your agency that cannot be accessed during high water, where rush hour traffic gridlock prohibits response, or do rail lines cross roadways that could potentially delay the arrival of cross-town companies?

Some jurisdictions have worked with the municipalities and highway departments to install traffic signal control devices on their apparatus (fig. 13–2). This relatively low-cost technology makes a tremendous improvement in run times, giving the approaching apparatus the right of way by changing a red signal to green as well as stopping oncoming traffic.

Fig. 13–2. Opticom™ infrared emitter shown on Metro West 3354. (Photo by Michael Digman)

One of the most influential factors on fire suppression is building construction. It is not uncommon for many of today's suburban fire agencies to be located in an area with a combination of new and old construction, and with varying aged roadways, water systems and, in some cases, limited or restricted access. Knowing the building construction your area will help you predict the fire behavior.

Perhaps the most important factor to consider when profiling the geography of your agency is the type of construction located within each segment. Are the buildings traditional type V combustible built in the post-war boom? Are they made with lightweight truss and joist systems? Do you have any actual type III structures in your area, a segment of balloon construction, or a country club community of McMansions? Each of these construction types brings differing variables to the success of the alarm directly proportionate to the amount of time the fire in the structure remains unchecked. Some agencies are arriving on the response scene 3, 5, or 10 minutes after the construction systems used for the building are primed to fail.

More than one of the NFPA and U.S. Census Bureau population density classifications can be found within the confines of an agency's borders. A response area composed of a combination of urban, suburban, and rural densities requires significant forethought and planning to accommodate the different challenges that lie within each type of population area. As you look at the descriptions of these classes, mentally apply them to the areas in your agency. Are you strictly suburban, or is your agency the more likely combination of several groupings? See figure 13–3.

Area Classifications

Metropolitan	Area of 200,000+ citizens, 3,000 people per square mile
Urban	30,000 citizens, 2,000 per square mile
Suburban	10,000 citizens, 1–2,000 per square mile
Rural	9,999 citizens, less than 999 per square mile

Fig. 13–3. Classifications of area based on population density.

Classifying an area with a more restrictive population designation results in a reduction in the NFPA standard response time. In some cases that drop is significant. In order to understand what your population demands, it is critical that you understand where the population is. By classifying your areas appropriately, you can adjust your agency's response package and future plans to meet the recommendations of the standards.

Areas that are composed of several density types must also maintain a fleet of equipment that is capable of protecting the construction, water supply, and infrastructure concerns associated with each.

For an agency to fully understand the response profile it is accountable for providing, it must also take a close look at the hazards that are transient within the area. Approximately 54% of the nation's hazardous materials are shipped over the road annually. While agencies might prepare for the manufacturers or processes that occur locally, knowing that an incident may occur with a product manufactured remotely and hauled through town increases the potential mitigation risks that the agency might be responsible for. Major interstates could bring any number of combinations of hazards that are not native to the jurisdiction. Rail and waterborne shipping, airports, and pipelines are also very real concerns.

While it is impossible to be prepared for every single emergency that might occur in your backyard, it is *critical* that you at least acknowledge the potential and plan for contingencies associated with the events.

Risk assessment profiling

As the fire service trends toward an *all-hazard response* profession, you should understand what you may be called on to mitigate and what you are *capable* of mitigating. Developing strategies, preplans, and specific fireground SOGs for a certain box or address will help in the overall process that is the reevaluation and future planning step of the suburban fire tactics assessment.

Large urban departments can detail a complete hazmat team. They can staff several technical and heavy rescue units. A joke in the FNDY is that they surround a structure fire with enough personnel that they can "simultaneously take a deep breath in, and deprive the fire of oxygen." This simply is not the case in smaller suburban fire departments and districts. Engines and ladders, squads, and rescues are often the same piece of apparatus rolled into one.

When planning for resource allocation and distribution on domestic and overseas bases, the United States Navy developed a risk assessment template (RAT). This standardizes the approach for assessing all structures on the base and determining a level of risk associated with each individual type of occupancy.

Obviously, a fueling hangar would be of substantially more risk than a barracks. By developing a standardized method of assessing risk, a baseline is determined and resulting tactical objectives for each level of risk are identified.

Metro West Fire District Structural Risk Assessment Template

Structure Address: ______________________________ Still Alarm Station:
City: ______________________________ 1 2 3 4 5
Occupancy Name: ______________________________
Building Dimensions L=______ W=______ #Floors ______ Date Assessed:
Shift/Company Completing: ______________________________ ____/____/____

Life Hazard: (normal occupancy)

High Life Hazard (100+)	3
Moderate Life Hazard (25–99)	2
Low Life Hazard (<25)	1

NFPA Construction Type

Type 2 (non-combustible)	3
Type 5 (lightweight wood frame)	2
Type 3 (ordinary)	2
Type 4 (heavy timber)	1
Type 1 (fire resistive)	1

Economic Impact

High	3
Moderate	2
Low	1

Water Supply within 800 feet

No municipal water	4
1 hydrant <1500 gpm (_____ft)	3
1 hydrant <1500 gpm (_____ft) *and* 1 750–1000 gpm (_____ft)	2
2 hydrants 1500 gpm+ (_____ft)	1

Building Usage

Industrial/commercial	3
Mercantile/small business	2
Residential	1

Square Footage

15,000 sq. ft. or greater	3
7501–14,999 sq. ft.	2
7500 sq. ft. or less	1

Access Issues

Limited (1–2 companies, 0 aerial)	3
Partial (aerial, 1–2 companies)	2
Unimpeded (aerial access, all companies)	1

Column Total ____

Setbacks

Greater than 300 ft.	3
100–299 ft.	2
Less than100 ft.	1

No water

No draft site available (1000 ft.+)	3
Draft site available (<1000 ft.)	1

Special Hazards

HazMat or explosives	3
Invalid/convalescent care	3
Heavily subdivided areas	3
Rack storage and/or small amount of HazMat or explosives	2
High-piled storage	2
Chemical process use	2
No special hazard	1

Column Total ____

Indicate if the following are present

Fire Protection

- ☐ Remote station monitored alarm
- ☐ Central station monitored alarm
- ☐ Sprinkler system
- ☐ FDC
- ☐ Areas of refuge
- ☐ Class 1 standpiping
- ☐ Class 3 standpiping
- ☐ Combustible exposures <25 ft.
- ☐ Transient use occupancy
- ☐ Structure warrants further evaluation

Forward to Capt. Digman upon completion

Fig. 13–4. A sample structural risk assessment form. (Illustration by Michael Digman)

This same practice can be applied to your jurisdiction. Assessing building construction, water supply, access issues, population, fire suppression systems, and other factors and then assigning a numerical risk value to these items generates a risk hazard level. Many times, companies take the risk assessment template with them as they conduct annual inspections. They reinforce protection systems, calculate fire flow, and review fire alarm locations.

Departments will find through the assessment of their agency that the classifications of buildings falls into predominantly two categories: average risk and high risk.

- Average risk structures are usually type V combustible, single-family dwellings. These are the structures that make up 99% of your emergency fire calls.
- Moderate/high risk structures vary in construction and occupancy, and differ from agency to agency. These buildings may have a particularly high life hazard, like a nursing home or multi-family dwelling with access issues, a structure used for hazardous materials storage or processes, or houses built in areas without access to public water service. Each jurisdiction will assess the risk level by its own standards and plan to that level.

Once the structures are assessed and plotted on the agency's map, real decisions can be made about asset allocation. Decisions include increasing or modifying the amount or type of apparatus on a response to this area in order to provide an appropriate protection for these buildings and accomplish the tasks that they deem necessary or critical. The process of developing tactical plans for these types of structures allows the leadership of an organization to plan for contingencies should an actual emergency occur.

As you perform the structural risk assessment it will become apparent where the high risk structures are in your agency. Are they close to or remote from your stations? Are they in areas where industry and residential life interfaces? What emergencies are likely in this area?

An interesting component to this risk assessment is the consequence component of the alarm. This refers to the economic or social significance of losing the occupancy to fire or other means. While this may or may not have direct correlation with the degree of life hazard, it is a unique component that is often overlooked by fire district planners.

While the loss of a single family residence or an entire apartment building is undoubtedly devastating to those families involved, there is little if any economic impact to the community in the way of lost revenue taxes by school and fire districts. Because most of these structures are insured, they are usually razed following a fire and quickly rebuilt. Fires in big box construction, schools, or offices often have the same effect, with services and jobs temporarily rerouted to other locations while the structure is repaired.

However, a fire in the only grocery or department store in a town, a manufacturing plant that employs half of the town's population, or other industry that provides significant financial or employment resources to the area can devastate a community. Often in these instances business owners may look to rebuild in the same location but find a cheaper site elsewhere or an alternative that provides a better opportunity for income. Businesses that choose not to rebuild or close permanently could devastate the community through the loss of jobs or tax revenue.

By applying this rubric (fig. 13–5) to the hazard classifications derived from the structural risk assessment by emergency type, you can begin to develop a feel and plan for the emergencies that are likely to occur in these areas. Take the opportunity to preplan the high-probability, high-consequence, maximum hazard emergencies and contingencies for each level of risk in between.

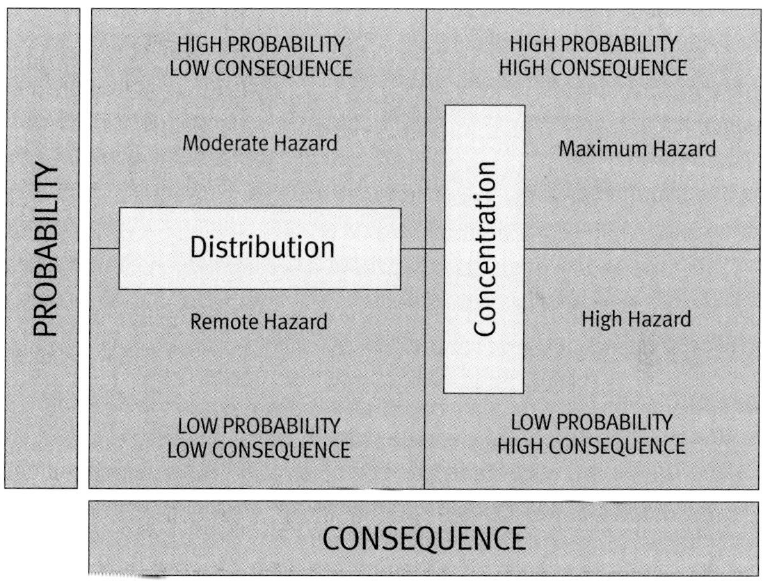

Fig. 13-5. A risk assessment rubric helps classify hazard level based on probability and severity. (Illustration by Michael Digman)

Defining probability and consequence helps to understand the impact and severity of potential incidents in the jurisdiction. Ultimately the priorities of our firefighting mission remain the same regardless of the structure that we enter to fight fire: life safety, property conservation, and confinement/extinguishment. We will do our best to accomplish these goals. It is incumbent upon fire agency leadership to account for these high-risk/high-consequence structures in their strategic plan, however, to ensure that any specialized needs, additional units detailed to the response, and contingency plans in the event of an occurrence are addressed.

Historical trending

Historical trending is perhaps the single most effective way to identify what a fire agency is actually providing the citizens on a daily basis. Through evaluating trends such as emergency type, frequency, distribution, call processing, response time, arrival of effective response force (ERF), and time to the completion of critical tasks, a statistical representation will detail what recommended benchmarks are met, where the department falls short, and identify areas where improvement can be made.

Where does this information come from? It is collected from the NFIRS (National Fire Incident Reporting System) or firehouse reporting that accompanies each run that an engine, ambulance, or rescue unit makes. Each alarm report that is submitted on behalf of an agency paints a picture of what that agency is currently producing. The more accurate the times and incident locations, the more consistent the coding of the alarm type and the actions taken, the better resource the alarm data becomes. As many agencies transition into computer automated dispatching (CAD) and global positioning system (GPS) address identification, our data set becomes more and more accurate.

The data is there for interpretation. The information derived is priceless to the planning process as your fire agency moves into the future. The data may show that an additional station is needed to serve an area and you will have the evidence to validate that statement to the city planners. The data may show that your response times are slowest Monday through Friday between 9 and 11:00 a.m., which may be the time that your companies are training. Further analysis of the information might show that you receive very few requests for service between the hours of 1 and 3 p.m. Should consideration be given to rescheduling training to this block of time to keep your assets in service during the busier demand periods?

Reporting applications such as FIREHOUSE Software®, NFIRS, and others objectively evaluate the times associated with an alarm beginning at the second the emergency telecommunications operator answers the line.

NFPA 1221 identifies call processing times and benchmarks for emergency telecommunications operators to process calls and dispatch the appropriate equipment. For example, when our agency studied its run times a significant delay was noted in call processing when our communications center added a new component to the CAD process. Calls were delayed as much as 2 minutes while the caller was interviewed to initiate EMD (emergency medical dispatching) and the appropriate apparatus were selected. Through no fault of the communications operators, the new SOG for the agency was adversely affecting their performance as they were trying to improve their product. Identifying this unforeseen effect on the alarm processing through our data analysis, a new procedure was implemented to start an ambulance immediately upon receipt and confirmation of an EMS alarm and then add any supplementary equipment that the CAD interview might indicate are needed. The end result was a reduction in the overall response times in the area from the moment of the 911.

This is just one example of how call analysis can benefit your agency. If performed properly, you will be able to identify the times of day and week when your stations are the busiest, perhaps allowing for adjustment of scheduled activities that pull an engine from that area during that time. You will be able to identify when firefighters are the slowest responding from the station or arriving at the call. Objective analysis of these factors will certainly identify areas where service can improve, training could benefit, or additions of apparatus need to occur.

Statisticians love to manipulate numbers to find the meaning of the data. They can find trends, outliers, and very valuable information from a data set. Firefighters benefit from the value of the same information. It is important that the data entered into your reports is as accurate as possible. Factors such as the number of firefighters on an engine; benchmark times such as responding or on scene that are documented with hours, minutes, and seconds; and the response nature correctly coded will make your data set more accurate as a representation of your agency. National consensus standards for response time, travel time, and effective response force arrival are recorded in seconds. Quite literally, in the realm of evaluation and future planning, every second counts.

Most new CAD programs and communications agencies are moving into GPS tracking of equipment. The old method of box assignment of predetermined ladders and engines to an address is being replaced with a set number of engines, ladders, rescue, chief, and ambulance units, where the *closest* units to the scene at the time of dispatch are pulled into the alarm. This brings in the closest trucks and affords the most rapid response. Programs can account for speed limits on roads, any detours or road closures, and traffic patterns and decide on the appropriate units within milliseconds.

Additionally, the GPS of the address of the alarm is captured. This allows organizational planners the ability to precisely annotate where the alarm occurred within the jurisdiction. By capturing this information for an entire year of alarms, zones of demand begin to develop. Comparing this information against previous years' data, demand trends quickly become apparent.

There are countless software applications for manipulating your data to produce very valuable and informative reports for your agency. These programs will analyze the information in seconds and paint a tremendous picture that outlines exactly what your agency is doing. The more intricate the program that is being utilized, the more helpful the information gleaned from your run times can be. Unfortunately, the information provided is only as accurate as the data that we record. It cannot be stressed enough that accurate data entry is vital to the process of evaluating performance of a fire agency.

Measuring Current Agency Performance

Progressive fire service leadership already knows if their agency is underserving a population. We know what is needed; convincing the governmental departments that apportion monies to provide the needed additional funding is more difficult. This process of reevaluation and future planning is exactly the weapon that firefighters need to justify the requests for additional staff or equipment.

Strategic deployment of assets

It goes without saying that it makes sense for the agency to place assets where they will do the most good. As we discussed previously in this chapter, stations and equipment historically often end up where they do out of convenience rather than necessity. The location of Station 2 may have been the best location in 1974; today it is remote from the population centers or removed from the major travel infrastructure. Geo-plotting the location of your stations into one of the analysis software programs or onto a large map of your agency that shows commercial and residential developments will paint a picture of area coverage.

Looking at the nature types of the emergencies mitigated in the last three years, are your assets best positioned to address these concerns? Is an engine serving an area with several multi-story apartment buildings when perhaps a ladder is needed? Is a quint with 500 gallons first due in an area with no water when an engine with 1,500 gallons of water might be available in another part of the district?

Agencies that have reciprocal agreements for an asset, such as a ladder tower or water unit, will want to position that asset where it is mutually beneficial for both agencies. When specialized equipment is located, infrastructure is often overlooked. Placing a unit where it can get to a major thoroughfare and ultimately where it's needed is advantageous for all who have stake in the unit.

We are not mandating that you move equipment to succeed in the implementation of the tactics outlined in this book. We do, however, suggest that you consider adapting your response package to address any glaring concerns or shortcomings in the agency through future purchase and planning.

SOPs/SOGs for response

Each agency and organization has its own goals; the priorities of how *they* fight a fire or conduct a search. If three agencies respond into an alarm, there is potential for the officers of these units to be on separate pages as to which tactical priority comes next. The tactics suggested in this section help to identify and stress the importance of developing a standardization of these goals and operating methods or procedures. By incorporating these procedures into standard operating guidelines (SOGs) and training with the companies that will respond on their alarms, confusion is minimized and every person's efforts are maximized. We don't attempt to develop your agency's SOGs, but stress the importance of engine and truck company work, rescue components, EMS for victims *and* firefighters, and a strong and present command structure.

Agencies must identify a list of critical tasks for the successful completion of an assignment in their jurisdiction, given the building construction, mutual aid agreements, and resources available to them prior to dispatch. Following are some examples:

- Attack line deployment
- Water supply
- Forcible entry
- Ventilation
- Aerial positioning
- Primary search

- Backup lines
- Ground ladders
- Rapid intervention team (RIT)
- Utilities secured
- Command
- Safety
- EMS

It is imperative that an agency looks hard at the number of firefighters they are receiving on a first alarm assignment, identify the list of critical tasks, and determine if sufficient personnel are on the scene to realistically, effectively, and *safely* complete those operations. Often, leaders discover that they are understaffed, which comes as no surprise, and add companies to the run cards to bring their numbers in line with what is needed to accomplish the objectives of the alarm.

A company isn't needed solely for utility control, nor is one needed only to position the aerial. Once these tasks are completed, these firefighters reenter circulation and can be reassigned.

Effective response force (ERF)

The effective response force (ERF) is a predetermined number of personnel that need to arrive on the scene of an emergency to accomplish a set list of tasks. This number may change with the location of an emergency. A fire in an area with no water is obviously going to require the establishment of a water supply, thus increasing the number of required staff at the scene. Your agency's operational staff should determine what your magic numbers are, then adjust your run configurations to ensure that your response meets these needs. For example, when our agency proposed and ultimately adopted a set of suburban fire tactics, an automatic provision was put in place to add two additional engines (hose and facilitator companies) to the alarm when a backup line was placed into operation.

Statistically, the ERF has significance because this level of response is tracked to measure the benchmark compliance for an organization. Translated, this means the clock stops when the ERF that you determine for a hazard class or risk type arrives on scene. If your agency does not have the proper resources to meet these times, the data will reflect your deficiencies. Conversely, the data will validate the work and planning that your agency is putting forward to protect the public if it is performing to standard.

Placing numbers on the staffing of a box assignment may be new to your organization. The ERF needs to address the *minimum* potential personnel that an apparatus may bring, and the *maximum* needed to accomplish a task (fig. 13–6).

Fig. 13–6. An effective response force may require the dispatch of additional units when a master stream is put into operation. (Photo by 5280 Fire)

Let's look at the same task list from the previous section, apply generalized apparatus assignments, and assume the following staffing of firefighters: Engine A (4), Engine B (3), Engine C (3), Ladder A (4), Chief A (1), Medic A (2), for a total of 17 arriving firefighters on the box assignment.

- Attack line deployment: 3 firefighters (EA 3)
- Water supply: 1 firefighter (EB 2)
- Forcible entry: 2 firefighters (EA 1/LA 1)
- Ventilation: 2 firefighters (LA 2)
- Aerial positioning: 1 firefighter (LA 1)
- Primary search: 2 firefighters (EB 2)
- Backup lines: 2 firefighters (EC 2)
- Ground ladders: 2 firefighters
- Rapid intervention team (RIT): 3-4 firefighters

- Utilities secured: 1 firefighter
- Command: 1 firefighter (CA 1)
- Safety: 1 firefighter
- EMS: 2 firefighters (MA 2)

You can see that several assignments/tasks are vacant given this alarm assignment. While several of the tasks that have yet to be assigned can be performed by one company, others such as RIT require a dedicated company for the duration of the alarm. It was this evaluation that led to the addition of an engine and command officer in our area to accomplish our identified scene tasks. By processing these numbers and tasks, your planning groups will see what if any additions are appropriate for your agency.

Call volume and alarm trending

Statistically, the fire department is a very predictable organization. Evaluation of the run data that is entered into the agency's reporting or records management programs will give a clear picture of where the agency is requested to go, what they are requested to do, and with what frequency.

As the data is manipulated with these programs, the results can be inserted into other programs to generate graphical displays of the information. These visual depictions are nice additions that bring the data to life during planning meetings. Areas that are over- or underserved are quickly identified, as are areas where responses are delayed, annual requests for service are increasing, and where the alarm types are trending.

With the aid of these visual presentations, processed alarms can be analyzed for the following components:

- Call processing time: How much time elapsed from 911 answer to alerting the engine/LSV?
- Turnout time: How much time elapsed from the alert of the engine to the response of that unit?
- Travel time: How much time did it take for the engine to arrive on scene?
- Arrival of ERF: What is the amount of time from initial dispatch to arrival of identified manning for ERF?
- Call density: What is the location of the majority of the alarms in the agency and how do they correlate to station/ apparatus allocation?

- Call trends: Is a specific location generating an unusual amount of alarms? When are they initiated and what are they for? Can this be attributed to a particular event at the complex, or are these alarms trending to be the new average for this location? Is there enough demand that temporary assignment of a unit to that area during these times might be warranted?
- Is the still alarm area of Station 4 exceptionally busy during the day because of workers that are in the area, but very slow at night? Is there merit in moving a unit to assist the Station 4 crew during these periods?

Ultimately, as the data is compiled and reviewed, the main focus of the analysis is whether or not the companies are making the response runs at or better than the NFPA recommended baseline travel times for that particular area's population density. If the times are not being met, a comprehensive review of the agency's operations is warranted to identify the shortcoming.

NFPA baselines and benchmarking

The National Fire Protection Association developed the consensus standard *NFPA 1710* in 2009. This standard essentially outlines suggested staffing levels for arriving apparatus at the scene of an emergency. While several emergency types are addressed, the standard predominantly refers to the personnel detailed to the scene of a working structure fire. These recommendations are labeled *recommended staffing* levels, and although they are part of a national consensus standard, there is no binding legal obligation for any agency to adopt them. Perhaps more important than the simple concept of a predetermined number of firefighters assigned to an engine company is the total number of firefighters needed to accomplish the predetermined tasks and on-scene benchmarks.

The *NFPA 1710* standard for structural firefighting recommends "two in, two out" as the standard practice of firefighters. Realistically, how does this standard apply to your organization? There are few suburban and rural departments that can meet the demanding requirements that this standard identifies on the initial arriving company of a first alarm assignment. While the FDNY, City of Miami, and other large urban area agencies may respond to the call with a "two in, two out" component on each engine, very few of the jurisdictions in my area can afford this level of staffing on a daily basis for an engine company. One nearby combination agency responds with one or two firefighters on the apparatus while other volunteer firefighters respond directly to the scene in their personal vehicles.

In addition to the staffing component of *NFPA 1710*, the standard identifies baseline and benchmark time limits as recommendations for responses. The recommended times are for both first due units as well as the ERF balance of the first alarm assignment. These times have an adjustment component depending on the type of area that an agency is serving. To hold a suburban agency to the same standard as a big city metropolitan area is not practical. The times need to be adjusted if infrastructure is not well developed, buildings are more spread out than in the urban or city setting, or the municipal water supply is not intact.

The response area demographics of building types and populations are used to help classify the area into zones. The local tax assessing agency can provide a count of the commercial, residential, and assembly properties in the predetermined areas of your district. The jurisdiction can also be divided using the subdivisions of U.S. Census tracts as a starting point. The Census and taxing agency should also be able to help identify the average persons per household (PPH) in that area. Numbers for each zone can be calculated by multiplying the average PPH by the number of residential structures in the area. Utilizing the classifications in figure 13–7, a classification of that zone can be determined.

NFPA 1710 Baseline Times

Call Processing Time	60 seconds
Turnout Time	80 seconds

NFPA 1710 Baseline Recommendations

Urban

For 90% of moderate fire responses
First arriving with four firefighters in 492 seconds (8:12 mins) total response time
ERF shall arrive within 804 seconds (13:24 mins) total response time

Suburban

For 90% of moderate fire responses
First arriving with four firefighters in 570 seconds (9:30 mins) total response time
ERF shall arrive within 960 seconds (16:00 mins) total response time

Rural

For 90% of moderate fire responses
First arriving with four firefighters in 960 seconds (16:00 mins) total response time
ERF shall arrive within 1272 seconds (21:12 mins) total response time

Fig. 13–7. *NFPA 1710* baseline times and recommendations.

These classifications are critical when applying the *NFPA 1710* recommended times to the assessment of your response performance. Misclassification of a zone can result in performance statistics that are grossly over or under the recommended benchmark times, thus skewing your performance accuracy.

Additionally, misclassification of an alarm type, response parameter, or inaccurate times in the documentation can skew the data set. An agency will have more than one classification of density within its borders. It is imperative that they be accurately identified so that a true measure of performance can be completed.

You might assume that these times are high. Keep in mind that these are the *minimum* recommendations for your agency's response. The point of the illustration is to note the vast difference in the times that adjust by the classification.

The baseline is the minimum standard; the benchmark is identified as the goal that the organization should strive for. You will note that these times are more restrictive. Also note that these goals should be moderately unattainable. That is to say, if you are meeting your benchmarks more than 90% of the time, you should adjust the benchmarks up to a more restrictive time frame. The idea is that you are consistently trying to better the product that your agency is putting on the street.

As a general rule, the more accurate the data you can analyze, the more accurate the overall snapshot of performance. Three years of alarm data is the accepted norm when calculating response times. It is important, however, that the data be recorded the same and the same parameters identified.

NFPA 1710 benchmark recommendations

NFPA/agency defined critical tasks. To evaluate established tactics and strategies, information must be targeted that pinpoints actions we can control and change. Fire service leaders must identify these individual actions and develop reporting systems to track data that fairly documents this accomplishment. Benchmarks, or critical event recognition, should be evaluated at every incident that requires service delivery (fig. 13–8).

Urban
For 90% of moderate fire responses
First arriving with four firefighters in 380 seconds (6:20 mins) total response time
ERF shall arrive within 620 seconds (10:20 mins) total response time

Suburban
For 90% of moderate fire responses
First arriving with four firefighters in 440 seconds (7:20 mins) total response time
ERF shall arrive within 740 seconds (12:20 mins) total response time

Rural
For 90% of moderate fire responses
First arriving with four firefighters in 740 seconds (12:20 mins) total response time
ERF shall arrive within 980 seconds (16:20 mins) total response time

Fig. 13–8. *NFPA 1710* benchmark recommendations.

Following are examples of NFPA fireground benchmarks:

- Response times: Times that are standardized for the response of the appropriate number of personnel to the scene.
- Staffing available: The effective response force. It should be noted that the standard calls for four firefighters on the first due company. This has been interpreted to include first arriving engine *and* support units, provided the company that is detailed to the engine remains assigned to the engine for the duration of the alarm and under direction of the company officer of the engine.
- Time for initial attack line deployment: The time it should take, in seconds from arrival, to stretch the line to the area of operation.
- Time for established water supply: The time it should take, in seconds from the arrival of the first due, to establish water supply.
- NFPA compliance standards
- Area of origin containment
- Documented rescues
- Fire fatality records
- Emergency responder casualty reports
- Reported Mayday documentation
- Other agency defined critical tasks

After-action reports

When the concept of suburban fire tactics and the "1 and 1" (one engine and one facilitating company) approach philosophy were adopted by agencies in our area, it was determined critical that training, review, and post-incident feedback be captured. This feedback is used to determine how the process is working, address any areas where deficiencies exist or refinement is needed, and reinforce the process so that it is second nature going forward. This reevaluation and reinforcement process works and the concepts are now second nature.

There are several means by which this evaluation can be accomplished in your organization. Simple critiques and coffee table discussions about the alarms are excellent ways to reinforce important points with your crew. Casual discussion that involves everyone that was on the run is the best way to determine why things were done the way they were, understand the thought processes of others, and to judge a level of comprehension of your tactical plan. It is important to note that these evaluations should be conducted in a non-confrontational fashion; shortly after the alarm so that the details are fresh, but removed enough from the emotions of the alarm so that there are no hard feelings if deficiencies are identified. Remember, the purpose is to better the response.

Conducting review and critiques of an alarm can be done in layers. Seize the opportunities to reinforce the objectives and better the performance of your crew by "striking while the iron is hot."

- An informal review of the alarm performance at the company level can occur on the front bumper of the engine while grabbing a bottle of water and a new SCBA (self contained breathing apparatus). Explain why tactical decisions were made and what occurred as a result of those decisions.

- A *hot wash* occurs at the completion of the alarm, usually back at the station, but not necessarily. Companies have the opportunity to explain their actions, describe the thought processes that went into the decisions they made, and give all members an opportunity to identify what they thought went well and what needs to be addressed in the future. It is important to note the term "hot" and understand that emotions are raw at this point. Guide members to keep things positive and nonconfrontational. This is not the forum to address a perceived mistake or shortcoming of another company. This is a forum for all companies to speak and review the call from their own perspectives. Emphasis should be placed on listening.

- A *call critique* occurs on a different duty tour shortly after the alarm. This is a great opportunity to involve all of the staff that responded to the alarm, including police agencies, EMS crews, fire marshals, and others. This is an organized debriefing where a single moderator (usually the command officer from the alarm) systematically goes through the alarm and reviews the call by order of arrival. This gives all in the room the opportunity to understand the thought process of the other companies, to hear what others did on the alarm and why they did so, and learn from the experiences of their counterparts. By including mutual aid companies, police, other agencies who were on the call, a very detailed review of the call is possible, thus maximizing the benefit and learning environment. This is a perfect place to reinforce the "1 and 1" principle, address why a company may have had to tactically go off script to address an issue, et cetera. Visual aids such as diagrams outlining where trucks were positioned, lines were laid, and hydrants were located will help a company who operated on the C exposure or the roof better understand what occurred on the A side of the alarm.

- A critical incident stress debriefing (CISD) is best addressed in a separate forum than those identified above. These debriefings are sometimes intensely emotional and, to be effective, often need to occur in a smaller group than outlined in above methods. That said, the formalized debrief or call critique is important to fact finding regarding the alarm so that a basis for later discussion can be established. A CISD should include anyone who *wants* to attend and should be offered to all. What constitutes an alarm that warrants CISD is unique to each individual. Nonetheless, valuable information often presents itself that is part of the reevaluation process when these debriefings are conducted.

Evaluation of Collected Data for Analysis

Once your organization's response data has been compiled, it needs to be manipulated so that it can better portray the information you need to appropriately determine the performance levels of your organization. The process of collecting and calculating this data can be done with simple spreadsheets or databases, but for those agencies with several thousand responses worth of data the task can be made more manageable by using a commercial software application.

There are a number of analysis applications available, and choosing the right one for your agency will depend on your evaluation needs and budgetary requirements. Whatever the means applied to your data set, make sure that it is reliable, reproducible, and understandable to the personnel that are utilizing it.

Statistical analysis of run data

The data will represent itself. There is little manipulation that can occur once the data is recorded. To ensure that the data produced is as accurate as possible, encourage those producing reports to complete them accurately and consistently. Train employees on what is expected in the report, how to classify runs, and which actions taken and representative times are required.

Once the data is compiled, you can begin to evaluate the data for input errors and validity. If you see an alarm that is grossly deviated from the remainder of the data set, a 24 hour response for example, evaluate the circumstances of the entry. It is possible that a missed entry for a date or typo was overlooked and can be easily corrected.

If there are calls that remain outside the norm, they should be excluded as statistical outliers. To include these calls will skew the data set and misrepresent how your agency is performing as a whole. An example of this skew occurs when *walk-in* alarms occur, where a citizen seeks help at the station and there is no response time. By including these alarms you would falsely reduce the average response time.

As your data is manipulated, reference it to the baseline and benchmark parameters established for the areas you are evaluating. Data plotting your alarms in the established hazard planning areas will allow you to apply differing response times to specific population densities.

The spreadsheet shown in the following illustration breaks down benchmark compliance for urban area population densities (fig. 13–9). It is further broken down into time of day in 6 hour increments, further examining the performance of the station during that period. Finally, the spreadsheet shows how often a station meets the urban benchmark on average.

It also illustrates the time needed to arrive at the scene of the alarms at 90% consistency. You will see in this example that Station 5 needs 615 seconds (10 minutes, 15 seconds) to arrive at a 90% average. What appears like poor performance is actually easily explained in that the station in question is in a rural setting, and the infrastructure is prohibitive of them making times at an urban benchmark.

Many records management applications will allow the user to plot data on a map. This map gives a visual representation of what the agency is doing and is a good way to display trends in an easy to understand manner (fig. 13–10).

Station	Bldg Fires	AM1	AM2	PM1	PM2	% @380 Secs	Secs to 90%	Secs to 80%	Median (secs)	Avg (secs)
4	20	73.04% (167)	92.27% (427)	91.26% (538)	91.12% (395)	90.26% (1,527)	385	330	239 secs	260 secs
1	18	82.65% (173)	89.68% (377)	89.46% (408)	89.17% (351)	88.81% (1,309)	395	345	262 secs	269 secs
3	19	63.15% (95)	85.56% (194)	81.02% (253)	82.91% (199)	82.32% (741)	440	380	294 secs	309 secs
2	5	58.57% (70)	81.08% (148)	76.55% (209)	70.62% (160)	74.44% (587)	520	420	300 secs	325 secs
5	14	30% (20)	70.17% (57)	59.61% (52)	46.16% (31)	57.23% (160)	615	530	355 secs	376 secs

Fig. 13–9. A sample benchmark compliance for an urban area.

Fig. 13-10. Plot data on a map to display trends.

If your records management software has a geo-plotting option, ensure that your alarm centers capture the required information and add that consistently to each alarm's report. It will only enhance the validity of the alarm data. Figure 13–10 shows the Metro West Fire District in St. Louis County, Missouri. The district is shown separated into equally sized grids with alarms for those grids tracked by type. An example is evident in G3. As the alarm density increased, the color escalated. Each of these grids can be set to a different population density standard for response time calculations. Visual aids such as these are helpful in presenting the findings of your analysis and identifying areas of concern.

Geographic/demographic response characteristics

The analysis of your data will show you exactly where your units are requested, the frequency that they are requested, and the distance from your stations your units are called to travel as they respond. The use of a geographic information system (GIS) analyzer to plot this data turns your numbers into visuals, better illustrating your responses (fig. 13–11).

Fig. 13–11. A sample GIS analysis to display call volume and trends. (Illustration by Michael Digman)

This report style is best run with one year's worth of data; more than that may skew the trend because of factors that change, such as construction on a major thoroughfare or a business that may have relocated out of that area. Using the most recent year's data will show you current trends. There is value in comparing the results to those of years past, but in summary form only. Adding that type of data may dilute your set and make the current trend analysis inaccurate. This screenshot from the NFIRS5 website (http://nfirs5.com) shows the response density for the grids established in an organization. Firehouse Analytics and Vinelight Fire Intelligence are other incident tracking reporting software programs.

NFPA 1710 compliance

Is your department compliant with the recommendations of the standard? Are you delivering enough personnel to the scene of the fire to accomplish the critical tasks outlined in *NFPA 1710*? More importantly, are you delivering the personnel to accomplish the tasks that you've identified as critical to the successful mitigation of a fire?

Often through review we determine that we are running short of staff on alarms. For example, you may find that tasks are not being handled properly due to insufficient personnel responding in on the initial dispatch. While it may be impractical from a budgetary standpoint to hire an additional firefighter for each engine company in your agency, this data analysis should be enough to augment your case to have additional companies from outside the agency added to your box assignments. Remember, equipment can be held in staging and easily returned if not needed. When it is needed, the travel time from quarters to the scene is fixed: you can't get it there any faster. The sooner you start them, the sooner extra hands arrive.

Consider adding another engine or ladder company for RIT (rapid intervention team) work. RIT is one of the most important, non-glamorous assignments on the fireground. It is critical that we have a crew in place to protect us and of sufficient size or number to handle the size of the occupancies. If that RIT crew is critically needed for another assignment, replace it with another company. Remember the "1 and 1" concept and to appropriately staff the alarm with extra companies as additional lines go into operation.

Are all populations adequately served?

There are certainly worse problems for an agency to have that that of overserving a population. If you have a station that is making metropolitan run time benchmarks in a suburban setting, congratulations are due. You have done an exceptional job of readying your equipment, preparing your infrastructure, and positioning staff and apparatus. Look at the health of your agency overall and assess if you have any voids that you can address.

Underserving population areas is a more likely result of your data analysis. While it can be alarming to find that there are areas that are not as protected as we initially thought they were, take an opportunity to address any issues that prevent your agency from meeting the expectations in that area. Are the issues temporary, such as road closures or construction? Was there a station closure, apparatus issue, or problem with mutual aid?

Measured Reactions to Data

Once the compilation of the statistical data is complete, a very distinct and irrefutable snapshot of the performance of your agency has been created. It makes sense to pay attention to what your data is telling you.

This data will serve you well as you proceed to the future planning stage for your organization. Where voids exist, begin the steps to address them. Where deficiencies prevail, look to leadership to help determine a solution. Ultimately it is our sworn duty to protect the citizens that have entrusted us with their safety. It may not be practical, or prudent, to go out and relocate a station simply because of what your statistics are illustrating. However, planning to address the deficiency is exactly what the intention of the reevaluation of your agency requires.

Are assets properly allocated?

By properly typing your requests for service, a clear picture of the breakdown of alarm type should be evident. Nationally, our service is trending so that the majority of our runs are EMS in nature. Your agency is likely no different. From there, miscellaneous *skills* such as responses for odors, cooking fires, code violations, and alarms sounding round out the more frequent requests for service. Finally, building fires complete the response profile.

Depending on the composition of your agency, you may or may not have an advanced life support (ALS) capable staff on board your engines. Just as in the baseline/benchmarks for moderate risk fire events, NFPA has established standard response times for EMS. Simply adding a paramedic with ALS equipment to an engine satisfies the requirements and thus meets the intent of the standard.

The GIS information of your data may show that the still alarm area for Station 4 is twice as busy as others between the hours of 0800 and 1800. Adding another unit during this time period, or moving another unit into the area to handle the increased load, might help alleviate the burden on your agency and improve your service delivery.

Your data may reveal that a specialized unit is better located in a station more central to your agency. Finding space for that unit in the apparatus bay is another conversation, but identifying that the unit is better serving the agency and the public and making preliminary plans to place it where it best meets the needs is exactly what this process is designed to do.

Examine response characteristics to find areas of improvement

After the data has been compiled, it can be broken down into the major classifications of measure:

- Alarm processing: Is call processing outside the normal limits? Address training issues and procedural compliance with your alarm agency. Find out if there is a way to streamline what they do prior to alerting units of the emergency to get help started.
- Turnout time: How long is the equipment taking to get out of the station? This may be as easy as a simple reminder to pick up the pace. Other measures to account for preparedness may be implemented to help get the unit out the door faster (fig. 13–12).
- Travel time: There is no easy way to decrease the amount of time that it takes to get to our alarms. As discussed, some agencies have installed traffic control devices to enhance their movement through traffic. Others have transitioned to a GPS-based CAD dispatch that pulls the closest units by accounting for speed limits, traffic patterns, and road closures. Training apparatus operators to utilize safe driving practices while choosing better routes are all ways to help speed the arrival of our units on scene.
- Effective response force (ERF) arrival: Again, speeding up the arrival of units on scene is not an easy process. Some of the ERF on your alarm may include the addition of a mutual aid company. If there is a different dispatching agency, there might be a delay in getting these assets to the scene. Creating a cooperative dispatching plan may help speed this process.
- Critical task completion (CTC): The hardest of the times to validate, CTC is important to measure but difficult to document.

There is no hard and fast way to track times of ventilation, water supply, completion of a primary search, or deployment and operation of the initial attack line. Training on these tasks will undoubtedly decrease the time it takes to perform them. Developing a comprehensive fireground SOG will further speed the accomplishment of these tasks, as they may be predesignated and therefore not have to be assigned.

It is in this area where the lowest cost, most influential, and simplest to implement changes can be made. Objectively evaluating these times will identify deficiencies. Proactive approaches to these deficiencies will improve the overall service to your community and assist your firefighters in mitigating their emergencies.

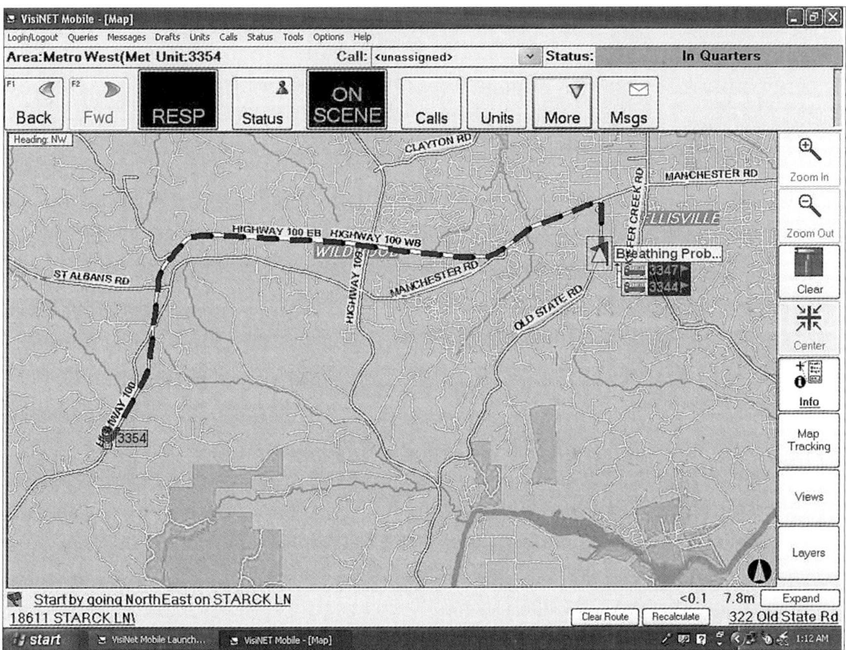

Fig. 13–12. A sample mobile data terminal (MDT) display showing a recommended route of travel to a medical emergency. (Illustration by Michael Digman)

Planning meetings and budgetary justifications

As we stated in the beginning, the fire service is being tasked with more and asked to do it with less. Today's financial times are not ideal for tax supported governmental agencies. As everyone tightens their belts, history shows this is exactly the time when we get busier and require more funding to support our operations.

Through a thorough reevaluation of your agency's compliance to national performance standards you will have a firm understanding of how your organization is positioned for success. While unrealistic to expect that identifying the need for four additional firefighters on the scene of a fire will result in the immediate hiring of new employees, it is realistic to expect organizational leadership to listen to what your data is telling you.

Through the interpretation of this information, you will know if you are making run times. You will know the busiest times of day and the busiest sections of town and be able to identify trends in requests for service. Armed with that information, you can submit budgetary requests for equipment, training programs, public education, outreach programs, staffing increases, and more. You will have hard and fast data to justify spending and asset or personnel allocation and to approach the public for increased budgets to support our common goal. The goal is simple: prepare the organization to protect the public.

Utilizing Information to Plan for the Future

Ultimately, the priorities of the governance and the leadership of the fire service organization are the same: provide life safety and property conservation services to the public in the most financially responsible means. How we meet these objectives is driven by the data that we develop through the examination of our alarm reports and the interpretation of that data by key decision makers.

Utilizing these facts to develop strategies to better your services is the primary goal of the reevaluation and future planning process. Take time to develop an understanding of what your agency is currently doing and compare that to the capabilities that exist in your organization. If there are simple improvements that can be made, do so immediately.

If the improvements are more drastic, such as relocating a fire station, adding staff, purchasing a specialized piece of apparatus, or shuttering a station and relocating the assets, the data is there to support your decisions. While these adjustments to the overall picture of your agency may not be immediate, they can be accounted for in your long-range and capital expenditure planning.

The use of this information to better your agency will ultimately improve the environment that your firefighters work in day in and day out. Positioning them for success will result in a better delivery of service to the public.

Index

B

C

D

E

F

G

K

L

M

N

O

P

Q

R

S

T

U

V

W

Z